# The RSGB RIG GUIDE

Edited by
**Malcolm Taylor**
G0UCX

Novice Notes by
**Robert Snary**
G4OBE

Published by the Radio Society of Great Britain
Lambda House, Cranborne Road
Potters Bar, Herts, EN6 3JE

© Radio Society of Great Britain 1996

All rights reserved. No part of this publication may be reproduced, stored in a retrieval system or transmitted in any form or by any means, electronic, mechanical, photocopying, recording or otherwise, without prior written permission of the Radio Society of Great Britain.

ISBN 1 872309 39 9

Design and typesetting by Malcolm Taylor Associates
Ashley Business Centre, Briggs House, Poole, Dorset BH14 0JR
Film output by Appletone Graphics, Moordown, Bournemouth,
Printing by Black Bear Press Limited, Cambridge

# Introduction

Welcome to the First Edition of the RSGB Rig Guide. I hope you enjoy browsing the pages and find it a useful addition to your shack, for discussing rigs during your QSO's, or for helping to identify the rig you want to own.

Some notes on the First Edition. The Guide's purpose to help you identify the key features of most of the transceivers available in the United Kingdom today. I use the term "most" deliberately. I found (or was told) that the number of pages needed to cover all rigs dating back only ten years, and avoiding one line list presentation, would put the book well outside the description of a "pocket" size reference guide! I also soon realised that the source of pictures suitable for reproduction ranged from excellent to non existent. For that reason, and at this point in the book, I make an early apology for some of the pictures.

The Guide is arranged as follows:

By manufacturer (alphabetically listed)

- New HF transceivers
- Used HF transceivers
- New VHF/UHF transceivers
- Used VHF/UHF transceivers

The current range of rigs are given more prominence, with the older ones sometimes being listed, several to a page. Many of the rigs are selected as suitable for the Novice Licence holder, and are highlighted with Novice Notes throughout the Guide.

A number of recent RadCom reviews are included at the end of the Guide, together with an index of past reviews, and rigs contained in the Guide.

Good browsing, I hope you enjoy, and good luck on your QSO's.

73's de Malcolm G0UCX

## Acknowledgements

I would like to thank the following people for their help and assistance in making this first Rig Guide:

Mike Dennison, Brett Rider, and Marcia Brimson, who first proposed the idea.

Dennis Goodwin of ICOM UK Ltd, who lent me his very comprehensive range of brochures on ICOM transceivers and made early comments on the Guide's design.

Robert Snary, G4OBE, who researched and supplied the Novice Sections to the guide.

Peter Waters of Waters & Stanton who very kindly supplied me with picture files from the Waters & Stanton catalogue.

Martin Lynch of Martin Lynch & Son who commented on the Rig Guide features and descriptions.

Also thanks to Barry Cooper of Yaesu UK Limited and David Wilkins of Kenwood UK Limited.

# Contents

## AR-146/446

The AR-146 is a 2 meter FM mobile transceiver giving 50 watts output. The AR-446 is the 70cms equivalent with 35 watts

**Frequency range**

Receive
130-174MHz (AR-146)
430-440MHz
Transmit
144-146MHz (AR-146)
430-440MHz (AR-446)

**Modes**

FM

**Output power**

50W, 10W, 5W (AR-146)
35W, 10W, 5W (AR-446)

**Features**

40 memories plus calling channel
Scanning band, memory, programmable
Steps 5, 10, 12.5, 15, 20, 25, 50kHz
LCD display
Optional CTCSS unit

## Sender 450

NOVICE NOTES

70 cm. The set is a budget priced handheld but easy to use, the LCD can be illuminated at night. The battery packs & additional modules also fit the Rexon RL-402 and possibly some other sets as well.
Some sets in AD450 cases did come on to the market for non amateur use covering frequencies in the range 300-399 MHz. The frequency display on this set shows the frequency as 3xx.xxx MHz.

**Frequency range**

430.000-439.995 MHz
steps user selectable.

**Output power**

2 watts with 9 volts or 5 watts with 12 volts

## AT-400

NOVICE NOTES

70cm. The set is a budget priced handheld and is easy to use. The LCD can be illuminated at night. The battery packs & additional modules also fit the Rexon RL-402 and possibly some other sets as well.

**Frequency range**

430.000-439.995 MHz
steps user selectable.

**Output power**

2 Watts @ 9 volts or 5 Watts @ 12 volts

## AT-200

Budget priced 2 meter FM handheld

**Frequency range**

Receive
130-170MHz
Transmit
144-146MHz

**Output power**

2.5 watts, 5 watts with 12 volts

**Features**

Options include:
CTCSS module
Power cable for 12 volts
Speaker microphone
Leather case
Dash mount

## 2001

144MHz mobile transceiver. Can be used as a base station.

**Frequency range**

144.500 - 145.975MHz

**Modes**

FM

**Output power**

25 watts high, 5 watts low

**Features**

Low cost
25kHZ channel spacing
66 channels
British built

## 4001

70MHz FM transceiver mobile or home.

**Frequency range**

70.250 - 70.4875MHz

**Modes**

FM

**Output power**

25 watts high, 5 watts low

## 6001

NOVICE NOTES

50 MHz FM transceiver mobile/home. Very simple but effective set for 50 MHz FM with the unusual distinction of being made in Britain. Easy to use and an ideal set for newcomers. Can be fitted with CAIRO-8 connector for RAYNET use or a different PROM so that the set initialises on packet frequencies.

**Frequency range**

**Output power**

25 Watts High 5 Watts Low

## 7003

70cms FM transceiver mobile/home. Very simple but effective set.

**Frequency range**

432.5 - 434.975MHz

**Modes**

FM

**Output power**

3 watts (for 5w dc in)

## 7001

NOVICE NOTES

70 cm MHz FM transceiver mobile/home. Very simple but effective set for 70 cm's FM with the unusual distinction of being made in Britain. Easy to use and an ideal set for newcomers. Can be fitted with a CAIRO-8 Connector for RAYNET use or a different PROM so that the set initialises on packet frequencies

**Frequency range**

432.5-435 MHz

**Output power**

3 watts (for 5w dc in)

# DX-70

Mobile HF and 6 meter transceiver. 100 watts output with detachable front panel (with extension cable kit).

**Frequency range**

| | |
|---|---|
| Receive: | 150kHz - 30MHz and 50-54MHz |
| Transmit: | All HF bands between 1.8 and 29.70MHz and 6 meters |

**Modes:**

SSB, CW, AM, FM

**Output power:**

100 watts (SSB, CW, FM), AM 40 watts

**Features:**

100 memories
Split frequency
Speech processor
Separate antenna sockets for HF and 6 meters
IRT
Squelch
Narrow filters
Programmable functions

## DR-610E

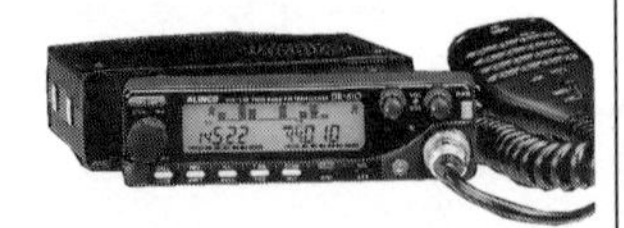

Dual band mobile transceiver with wide band AM and FM receive. Detachable front panel for mobile mounting

**Frequency range**

Receive:
108-174MHz FM/AM
400-510MHz FM/AM
830-990MHz FM
Transmit:
144 - 146MHz
430 - 440MHz

**Modes:**

FM, AM (receive only)

**Output power:**

50W, 10W, 5W VHF
35W, 10W, 4W UHF

**Features:**

Wide band receive
Channel scope
Mono, stereo
Detachable front panel (optional cable required)
Duplex operation
Remote DTMF control
Cross band repeater
RF attenuator
Separate sockets for VHF, UHF
Offset fully programmable

## DR-605E

Dual band VHF and UHF mobile transceiver.

**Frequency range**

144-146MHz
430-440MHz
Extended receive range

**Modes:**

FM

**Output power:**

50W VHF
35W UHF

**Features:**

Full duplex
100 memories
9600bps Packet socket
CTCSS encode
1750Hz tone
Programmable shift
Programmable steps
Low power position
Extended receive

## DR-130/430E

The DR-130 is a VHF mobile transceiver with the DR-430 being its UHF equivalent

**Frequency range**

144 - 146MHz
430 - 440MHz

**Modes:**

FM

**Output power:**

50W 5W (DR-130E)
35W, 5W (DR-430E)

**Features:**

Synthesised 20 channel capability with option to extend to 100 channels
20 memories (100 option available)
Programmable channel steps
Offset programmable
CALL channel button

## DR-150E

VHF mobile transceiver with VHF and UHF wideband AM, FM receive. Released 1996.

**Frequency range**

Receive:
108-174MHz
400-480MHz
800-950MHz
Transmit:
144 - 146MHz

**Modes:**

FM, AM (receive only)

**Output power:**

50W, 25W, 10W

**Features:**

100 memories
Channel scope
CTCSS encoder
DTMF encode and decode
Dual VFO
Search and scan function
Programmable shift
Reverse repeater
Receiver switchable AM, FM

## DR-M06/MO3

6 meter (DR-M06) or 10 meter (DR-M03) mobile transceiver.

**Frequency range**

50-54MHz (DR-M06)
28-29.7MHz (DR-M03)

**Modes:**

FM

**Output power:**

10W, 1W

**Features:**

100 memories
CTCSS encode installed (decode optional)

## DJ-G1E

2 meter handheld transceiver with wideband receive

**Frequency range**

Receive:
108 - 174MHz
400 - 510MHz
800-950MHz
AM, FM switchable
Transmit:
144 - 146MHz

**Modes:**

FM, AM receive only

**Output power**

2 watts. 5 watts with 12V DC supply

**Features:**

80 memories
3 adjacent channel monitor
Programmable channel steps
Airband receive
CTCSS encode
DTMF built-in
6 scan modes
Auto power off

ALINCO

VHF/UHF RANGE

## DJ-G5E

Dual band VHF/UHF handheld transceiver with extended receive on AM and FM

**Frequency range**

Receive:
108 -174MHz AM, FM
400 - 511.995MHz FM
800-999.990MHz FM
Transmit:
144 - 146MHz
430 - 440MHz

**Modes:**

FM, AM (receive only)

**Output power:**

Up to 5 watts

**Features:**

100 memories
CTCSS encode
DTMF
Channel scope
AM airband receive
Cross band repeater
Channel number display

## DJ-F4E

70cm 2 watt handheld transceiver. Ideal for the Novice licence holder.

**Frequency range**

Receive
430-470MHz
Transmit
430-440MHz

**Modes:**

FM

**Output power:**

2W, 1W or 100mW

**Features:**

40 memories
Options include:
Tone squelch
Headset with VOX
Up-Down speaker microphone

## DJ-180/480

VHF (DJ-180) or UHF (DJ-480) handheld transceivers

**Frequency range**

Receive:
130 - 170MHz (DJ-180)
400 - 520-MHz (DJ-480)
Transmit:
144 - 146MHz (DJ-180)
430 - 440MHz (DJ-480)

**Modes:**

FM

**Output power:**

2 watts. 5 watts with 12V DC supply

**Features:**

10 memories expandable with option to 50.
Scanning
Programmable steps
Rotary frequency control
Low battery indicator

## DJ-190

Compact VHF handheld transceiver with extended receive range.

**Frequency range**
Receive:
135 - 174MHz
Transmit:
144 - 146MHz
**Modes:**
FM
**Output power:**
Up to 5 watts with external supply
**Features:**
40 memories
Variable offset
5 - 30kHz steps
1750 Hz tone built-in
CTCSS encode

## DJ-191E

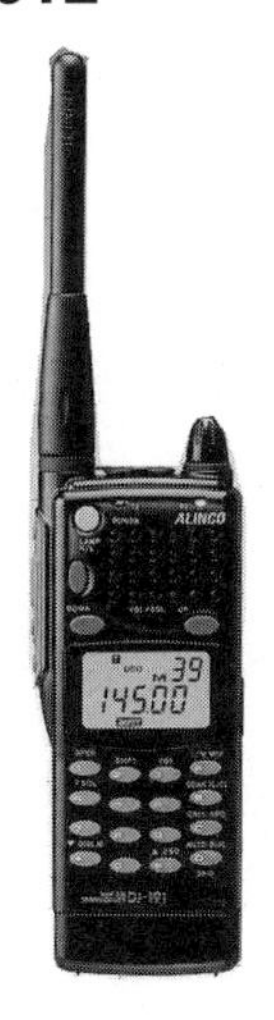


2 meter compact handheld transceiver

**Frequency range**
144 - 146MHz
**Modes:**
FM
**Output power:**
Up to 5 watts with external supply
**Features:**
40 memories + call channel
Variable offset
1750 Hz tone built-in
CTCSS encode
DTMF squelch encoder and decoder built-in
Scan functions
Options include:
CTCSS decoder
Speaker microphone
Headset with VOX

## DJ-S41

NOVICE NOTES

FM Mobile transceiver. The DR-430E is a very compact mobile transceiver which can be used at home provided a suitable PSU is used

**Frequency range**
430-440MHz
**Modes**
FM
**Output power**
35W, 5W

## DR-430E

NOVICE NOTES

FM mobile transceiver. The DR-430E is a very compact mobile transceiver which can be used at home provided a suitable PSU is used. It comes with built in CTCSS encode and has a very easily read display.

**Frequency range**
430-440mhz
**Output power**
35W, 5W

## DJ-F4E NOVICE NOTES

FM handheld. A small transceiver which can be used to great effect by novices. The keypad can be illuminated for night time ease of use.

**Frequency range**
430-440mhz
**Output power**
1.5W,1.0W,100mW (on battery)

## DJ-480EB NOVICE NOTES

FM handheld Transceiver. A fully functional handheld which is very easy to use and ideal for the newly licensed novice.

**Frequency range**
430-440mhz expandable
**Output power**
2w Battery - 5w 12 V D.C.

*Alinco DR-130 and DR-430 transceivers*

# IC-775/IC-775DSP

Base station HF transceiver released October 1995. Two versions available with (IC-775DSP) or without (IC-775) digital signal processing. Completely new design but logical step from the IC-765. DSP version adds enhanced noise reduction facilities, digital modulation and demodulation, automatic notch at audio frequency and digital filters.

**Frequency range**

| | |
|---|---|
| Receive: | 100kHZ - 29.990MHz |
| Transmit: | All HF bands between 1.8 and 29.70MHz |

**Modes:**

SSB, CW, AM, FM, RTTY

**Output power:**

5-200 watts (SSB, CW, RTTY, FM), 5-50 watts (AM)

**Features:**

Power supply included
Twin pass band tuning
CW Reverse mode
Manual notch
Auto notch (DSP version only)
1 Hz tuning
Dual watch (two in band simultaneous receivers)

# IC-756

First shown at Leicester during October 1996. Replaces the IC-736. Unique is a 4.9 inch LCD display which gives it a similar look to the IC-781 even though that rig used a CRT for its display. The display allows viewing of band scope, soft key functions and TX CW memory messages. The IC-756 transmits on HF and 6 meters with a general coverage receiver. DSP (Digital Signal Processing is standard.



**Frequency range**

| | |
|---|---|
| Receive: | 100kHZ - 29.990MHz |
| Transmit: | All HF bands between 1.8 and 29.7MHz plus 6 meters |

**Modes:**

SSB, CW, AM, FM, RTTY

**Output power:**

5-100 watts (SSB, CW, RTTY, FM), 5-50 watts (AM)

**Features:**

Twin pass band tuning
CW Reverse mode
Manual notch
Auto notch (DSP version only)
1 Hz tuning
Dual watch (two in band simultaneous receivers)

# IC-728

Compact and lightweight HF band transceiver. Requires external power supply. Conventional ICOM display and S meter. DDS (Direct Digital Synthesizer System) tuning. Options for AM and FM transmitting and receiving.

**Frequency range**

| | |
|---|---|
| Receive: | 30kHZ - 30MHz |
| Transmit: | All HF bands between 1.8 and 29.70MHz |

**Modes:**

SSB, CW (AM and FM if options fitted)

**Output power:**

10-100 watts SSB, CW (FM if fitted), 10-40 watts (AM if fitted)

**Features:**

Optional FM, AM board
Dual VFO's with split capability
Band stacking
30 memory channels
10Hz resolution

# IC-706

Released early 1995, and immediately impressed the world with its broad range of frequency and its physical size. ICOM UK had sold the 1000th unit by October 1996. The front panel can be detached which, with an optional cable allows the main unit to be housed away from the car dashboard.

**Frequency range**

| | |
|---|---|
| Receive: | 30kHZ - 200MHz |
| Transmit: | All HF bands between 1.8 and 29.70MHz, 6 meters, 2 meters |

**Modes:**

SSB, CW, AM, FM, RTTY (FSK), WFM (receive only)

**Output power:**

5-100 watts (All HF bands and 6 meters - SSB, CW, RTTY), 10 watts (2 meters - SSB, CW, FM)

**Features:**

Continuously receives 30kHz to 200MHz
1 Hz tuning
Optional antenna tuner for HF and 6 meters
Dot matrix display with spectrum scope
102 memories
Built in electronic keyer
IF shift

## IC-720

Released 1980 and retailing at £990.00 the IC-720 boasts 100 watts HF and a general coverage receiver.

Power output is provided by two 2SC2097 transistors in a wideband, push pull power amplifier.

**Frequency range**
Receive:
0.1MHz - 30MHz
Transmit:
All HF bands between 1.8 and 29.70MHz

**Modes:**
SSB, CW, AM, RTTY (FSK)

**Output power:**
200 watts PEP (SSB, CW input, 40 watts AM

**Features:**
10Hz tuning
Passband tuning
RF speech processor
VOX
Dual VFO's with split capability

## IC-725

Compact HF transceiver with general coverage receiver. Double conversion superheterodyne (triple on FM)

**Frequency range**
Receive:
30kHz - 33MHz
Transmit:
All HF bands between 1.8 and 29.70MHz

**Modes:**
SSB, CW, AM, FM

**Output power:**
10-100 watts (SSB, CW, FM*) 10-40 watts AM*

**Features:**
*Optional FM/AM board
Dual VFOs with split capability
Band stacking
30 memory channels
10 Hz resolution

## IC-726

Released 1989 and retailing at £989.00. HF and 50MHz transceiver with general coverage receiver and 50MHz.

**Frequency range**
Receive:
30kHz - 33MHz plus 46.20000 - 61.10000MHz
Transmit:
All HF bands between 1.8 and 29.70MHz and 50MHz

**Modes:**
SSB, CW, AM, FM

**Output power:**
10-100 watts (SSB, CW, FM) 10-40 watts AM

**Features:**
26 memory channels

## IC-729

HF and 50MHz transceiver with general coverage receiver. Compact and light, weighing only 10.8lbs. Receiver is a triple conversion superheterodyne

**Frequency range**

Receive:
30kHz - 33MHz
Transmit:
All HF bands between 1.8 and 29.70MHz plus 50MHz.

**Modes:**

SSB, CW, FM, AM

**Output power:**

10-100 watts SSB, CW, FM and 10-40 watts AM

**Features:**

Pre-amplifier
26 memory channels

## IC-730

Released 1985 and retailing at £659.00. Low cost (entry) HF rig. Compact measuring only 3.7in height, 9.5in width and 10.8in depth.

**Frequency range**

Receive:
All HF bands from 1.8MHz to 29.7Mhz
Transmit:
All HF bands between 1.8 and 29.70MHz *except* WARC bands

**Modes:**

SSB, CW, AM

**Output power:**

10-200 watts PEP (SSB, CW) 10-40 watts AM

**Features:**

Receiver pre-amp built in
Noise blanker
RIT
Speech processor
IF shift
Fully solid state
Digital read-out

## IC-735

HF transceiver with general coverage receiver.
PLL synthesizer tuning.

**Frequency range**

Receive:
0.1Mhz - 30.0MHz
Transmit:
All HF bands between 1.8 and 29.70MHz

**Modes:**

SSB, CW, AM, FM

**Output power:**

10-200 watts PEP (SSB, CW, FM) 10-40 watts AM

**Features:**

PLL synthesizer
6 digit illuminated LCD display
Built in receiver pre-amp
Matching ATU option
Requires external power supply

ICOM

HF RANGE

## IC-737

Compact HF transceiver with general coverage receiver.

**Frequency range**
Receive:
30kHz - 30MHz
Transmit:
All HF bands between 1.8 and 29.70MHz

**Modes:**
SSB, CW, AM, FM

**Output power:**
10-100 watts (SSB, CW, FM) 10-40 watts AM

**Features:**
Built in receive pre-amp
Requires external power supply
Triple conversion superhet receiver

## IC-740

HF transceiver with HF band (WARC bands included) coverage receiver.

**Frequency range**
Receive:
All HF bands between 1.8 and 29.70MHz
Transmit:
All HF bands between 1.8 and 29.70MHz

**Modes:**
SSB, CW, RTTY, (FM option)

**Output power:**
200 watts PEP

**Features:**
Options included are:
Marker unit
FM unit
Electronic keyer
Built in power supply

## IC-745

Base station HF transceiver and general coverage receiver.

**Frequency range**
Receive:
0.1MHz - 30MHz
Transmit:
All HF bands between 1.8 and 29.70MHz

**Modes:**
SSB, CW, FM (option), RTTY, (AM receive only)

**Output power:**
200 watts PEP SSB
200 watts input CW
10-100 watts

**Features:**
Options included are:
Marker unit
FM unit
Electronic keyer
Built in power supply
Filters

## IC-751

Base station HF transceiver with general coverage receiver. All modes.

**Frequency range**

Receive:
0.1MHz - 30MHz
Transmit:
All HF bands between 1.8 and 29.70MHz

**Modes:**

SSB, CW, AM, RTTY (FSK), FM

**Output power:**

200 watts PEP (SSB, CW input, 40 watts AM. Continuously adjustable from 10 - maximum watts.

**Features:**

Electronic keyer included
32 memories
Passband tuning
Notch filter
Speech compressor
Options include:
Internal power supply
Computer interface
Voice Synthesizer

## IC-761

Base station HF transceiver with general coverage receiver.

**Frequency range**

Receive:
0.1MHz - 30MHz
Transmit:
All HF bands between 1.8 and 29.70MHz

**Modes:**

SSB, CW, AM, RTTY, FM

**Output power:**

10-100 watts (SSB, CW, FM) 10-40 watts AM

**Features:**

Receiver pre-amp built in

## IC-765

Base station HF transceiver and general coverage receiver.

**Frequency range**

Receive:
0.1MHz - 30MHz
Transmit:
All HF bands between 1.8 and 29.70MHz

**Modes:**

SSB, CW, AM, RTTY, FM

**Output power:**

10-100 watts (SSB, CW, FM) 10-40 watts AM

**Features:**

Receiver pre-amp built in

## IC-781

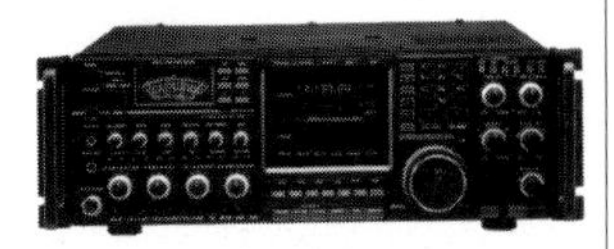

Base station HF top of the range transceiver. Released 1988 and retailing at £4500. Unique multi-function CRT display in centre of unit.

**Frequency range**

Receive:
0.1MHz - 30MHz
Transmit:
All HF bands between 1.8 and 29.70MHz

**Modes:**

SSB, CW, AM, RTTY (W), RTTY(N), FM

**Output power:**

150 watts PEP (SSB, CW input, 75 watts AM

**Features:**

10Hz or 1kHz tuning
Passband tuning
RF speech processor
VOX
Dual VFO's with split capability
Quadruple superheterodyne receiver (triple on FM)
CRT display

## IC-R71E

General coverage base station receiver with digital display and S meter. Continuous tuning from 0.1MHz to 30MHz

**Frequency range**

Receive:
0.1 - 30MHz

**Modes:**

SSB, CW, AM, RTTY, (FM optional)

**Features:**

32 memory channels
Optional remote controller
Programmed memory
IF notch filter
Noise blanker
Keyboard frequency entry
Built-in pre-amplifier and attenuator
Optional voice synthesizer

## IC-2GXE/IC4GXE

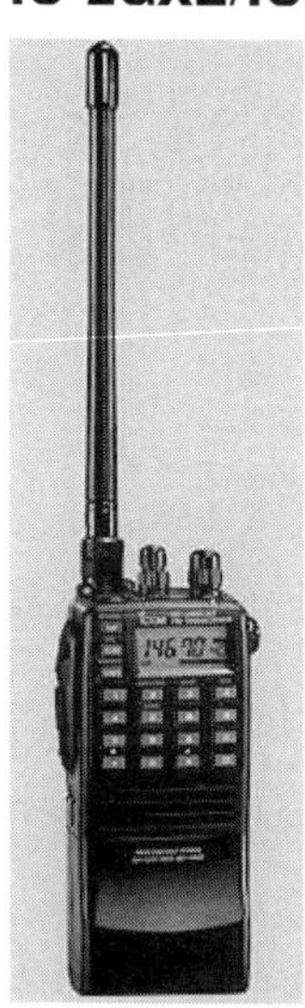

The IC-2GXE is a 2 meter handy and the IC-4GXE the UHF model. 7 watts output ((6 watts 4GXE)

**Frequency range**

Receive:
144 - 146MHz IC-2GXE
430 - 440MHz IC-4GXE
Transmit:
144 - 146MHz (IC-2GXE) or 430 - 440MHz (IC-4GXE)

**Modes:**

FM

**Output power:**

7 watts VHF model
6 watts UHF model

**Features:**

40 memory channels

## IC-T22E/ICT42E

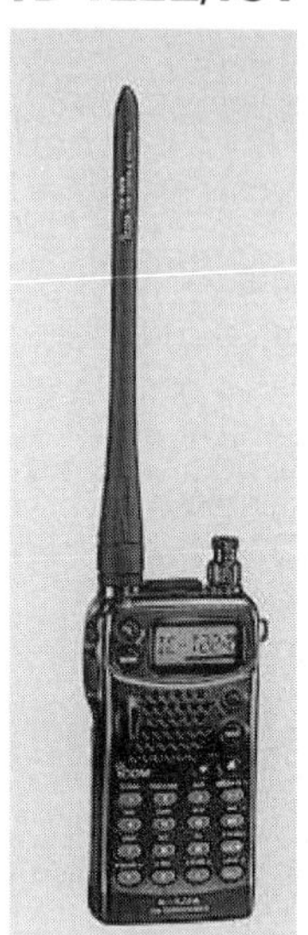

The IC-T22E is a popular 2 meter ultra compact handy and the IC-T42E the UHF model.

**Frequency range**

Receive:
144 - 146MHz IC-T22E
430 - 440MHz IC-T42E
Transmit:
144 - 146MHz or 430 - 440MHz depending on model

**Modes:**

FM

**Output power:**

5 watts high, 0.5 watts low (VHF), 3.5 watts high, 0.5 watts low (UHF)

**Features:**

Ultra compact size
Alphanumeric memory channels
Alphanumeric pager
CTCSS tone encoder option
Numerous scan functions

## IC-2000H

Mobile 2 meter FM transceiver. 50 watts output, 50 memory channels

**Frequency range**

Receive
144 - 146MHz
Transmit
144- 146 MHz

**Modes:**

FM

**Output power:**

High 50 watts, low 10 or 5 watts

**Features:**

50 memory channels
Optional message function

## IC-T7E

Popular dual-band (VHF - UHF) handie. Very small size. Released 1996.

**Frequency range**
Receive:
118 - 174MHz
400 - 470MHz
Transmit:
144 - 146MHz
430 - 440MHz

**Modes:**
FM

**Output power:**
4 watts - 0.5 watts VHF
3 watts - 0.5 watts UHF

**Features:**
Optional microphone
Simple remote system
9 DTMF memories
Numerous scan functions
Channel indication
CTCSS programmable

## IC-W32E

Dual-band (VHF - UHF) handie. Very small size. Released 1996.

**Frequency range**
Receive:
118 - 174MHz
400 - 470MHz
Transmit:
144 - 146MHz
430 - 440MHz

**Modes:**
FM

**Output power:**
5W, 0.5W, 15mW

**Features:**
Full cross band duplex
6 memories

## IC-2350H

Mobile dual band transceiver. High speed scans. 110 memories.

**Frequency range**
Receive:
118 - 174MHz
400 - 470MHz
Transmit:
144 - 146MHz
430 - 440MHz

**Modes:**
FM

**Output power:**
50W, 10W, 5W

**Features:**
Ultra high speed scans

## IC-2710H

Compact mobile VHF and UHF transceiver.

**Frequency range**
Receive:
118 - 174MHz
400 - 470MHz
Transmit:
144 - 146MHz
430 - 440MHz

**Modes:**
FM

**Output power:**
50W, 10W, 5W

**Features:**
Detachable front panel with optional separation cable
Full remote control microphone with backlighting

## IC-821H

Dual band VHF/UHF base station transceiver. Released October 1996 replacing IC-820H.

**Frequency range**
Receive:
136 - 174MHz
400 - 470MHz
Transmit:
144 - 146MHz
430 - 440MHz

**Modes:**
FM, SSB, CW, AM

**Output power:**
6W, 25W, 35W (SSB) on 144MHZ
6W, 40W, 30W (SSB) on 430MHz

**Features:**
Satellite VFO and 10 satellite memory channels
High sensitivity
1Hz tuning
DDS tuning system
RIT
IF shift
AF speech processor
CW semi break-in
Tone squelch
Optional tone scan
Supports 9600 bps packet

## IC-4E

NOVICE NOTES

70 CM fully synthesised transceiver. One of the first synthesised F.M transceivers. Extremely popular, and although lacking in memories still a very good performer. The set must not be run off voltages higher than 10.8 volts without the use of the IC-DC1. You may see battery packs or accessories marked as CP or CM these are commercial variants of amateur products and can be used.

**Frequency range**
430.000- 439.995 MHz in 5kHz steps.
**Output power**
1.5 Watts High

## IC-O4E

NOVICE NOTES

70cm fully synthesised FM transceiver. One of the first Keypad controlled transceivers, the LCD is small and although it can be illuminated can be difficult to read at night.

**Frequency range**
430. - 439.995 MHz steps selective by user.
**Output power**
1 Watt High on IC-BP3

## IC-12E

NOVICE NOTES

Fully synthesised FM transceiver. A 23 cm version of the IC-04E and one of the first 23 cm transceivers. Very similar in appearance to the IC-04E and similar comments as regards the display.

**Frequency range**
1240-1300 MHz steps selectable by user.
**Output power**
1 Watt High on IC-BP3

## IC-402S

NOVICE NOTES

70cm SSB/CW transceiver. When the set was first produced, it attracted a lot of interest from satellite operators as it had the facilities for both upper & lower sideband operations. It is quite a rare set to see on the secondhand market and is still sought after. Although SSB has advantages over FM as regards distance worked. There will be many times when there is a lack of activity in the SSB segment of the band

**Frequency range**
432.0 - 432.4 MHz plus two additional ranges
**Output power**
3 Watts PEP

## IC-30A

NOVICE NOTES

Crystal controlled FM mobile. This set was one of the first commercially available 70 cms FM mobiles and has a reputation as a reliable old work horse. The set was very similar to the IC 22A (for 2 metres) and can provide a reliable entry set. The main disadvantage is the set is crystal controlled. To get additional channels it may be nesssary to contact specialist firms to get crystals ground to order.

**Frequency range**
430-440 Mhz
**Output power**
10 W High 1 Watt Low

## IC-490E

NOVICE NOTES

Synthesised multimode mobile. The set has 5 memories, dual VFO's and also the facility to listen on the input when using repeaters.

**Frequency range**
430-440 MHz (steps according to mode )
**Output power**
10 Watts / 1 Watt

## IC-481H NOVICE NOTES

FM mobile transceiver. Dual band receive and full duplex (cross band) . The set also has a dedicated data connector for 9600 Baud packet operation.

**Frequency range**
430-440 MHz Transceive
144-146 MHz receive only

**Output power**
35 Watts / 10 Watts / 5 Watts

## IC-T42E NOVICE NOTES

FM portable transciever. The set is a very compact hand held, which is capable of having the memory channels labelled with alpha numeric designations. It is also capable of "paging" functions using the in built DTMF facilities.

**Frequency range**
430 - 440 Mhz

**Output power**
5 w / 0.5 w ( 13.5 v or 9.6 v supply ) 3.5 / 0.5 w ( 7.2 v supply )

## IC-4GXE (T) NOVICE NOTES

FM handheld transciever. Very durable set in two versions. The IC-4GXET has a key pad on the front panel and comes with CTCSS and DTMF facilities as standard. The IC-4GXE is not capable of DTMF operation and has the CTCSS module as an optional extra, and so could be regarded as the entry level set.

**Frequency range**
430 - 440 Mhz

**Output power**
6 W / 1 W ( 13.5 V D.C. supply )

## IC-P4E

FM handheld transceiver

## IC 505 NOVICE NOTES

SSB/CW transceiver. The set was very much a hybrid being in an uncertain area of being a portable and a mobile. In order to operate FM you needed to purchase an optional module.

**Frequency range**
50 - 54 Mhz

**Output power**
3 W / 0.5 W (BAT) 10 W / 3W / 0.5 W 13 .8 V

## IC 451E NOVICE NOTES

Multimode transceiver. A 70 cms multimode base transceiver which was one of the first to be introduced. Renowned for good audio and reasonably sensitive.

**Frequency range**
430 - 440 MHz

**Output power**
10 Max Variable

## IC 471 E (H) NOVICE NOTES

Multimode transceiver. A replacement for the IC451E and a good Multimode set. Quite sensitive but for top DX hunting, stations may require the mast head pre-amp to give the final edge.

**Frequency range**
430 - 440 MHz

**Output power**
25 Watts Max (Variable)
(75 Watts H version)

## IC 475 E (H)

NOVICE NOTES

Multimode transceiver. The IC 475 finished its production run in late 1995 after having been the successor to the IC471 and marked the end of a line in mono band multimode transceivers. Very sought after, particularly the "H" variants as they will run virtually all day at novice power levels without overheating.

**Frequency range**

430 - 440 Mhz

**Output power**

25 Watts max variable (H version 75 Watts)

## IC 551

NOVICE NOTES

Multimode 50MHz transceiver, which was available just prior to the band being released for UK operators. Some sets were imported from the USA as personal imports, although later sets were imported via UK distributors. The set operates CW, SSB & AM as standard, with FM being available with the purchase of an optional board.

**Frequency range**

50-54 MHz

**Output power**

80 W Max variable

## IC 571 E (H)

NOVICE NOTES

Multimode transceiver. A replacement for the IC551E and a good multimode set.

**Frequency range**

50 - 54 MHz

**Output power**

25 Watts max (variable) (75 Watts H version)

## IC 575 E (H)

NOVICE NOTES

Dual band multimode transceiver. The IC575 finished its production run after having been the successor to the IC571. Very sought after, particularly the "H" variants as they will run virtually all day at novice power levels without overheating. The set covers two bands which are of particular interest to class "A" Novices. 10 metres is under used during solar minimum, it would provide a lot of activity for Novices at the time of solar maximum.

**Frequency range**

28-30 Mhz & 50-54 Mhz

**Output power**

25 Watts Maximum Variable (H Version 75 Watts)

## IC 1271 E

NOVICE NOTES

Multimode transceiver. The set was introduced as possibly the first commercially multimode transceiver for 23 cms, it is quite a rare set to see on the second hand market.

**Frequency range**

1240- 1300 MHz

**Output power**

10w

## IC 1275 E

NOVICE NOTES

Multimode transceiver. The IC 1275 was the successor to the IC 1271.

**Frequency range**

1240-1300MHz

**Output power**

25Watts Max Variable

## IC-47E

NOVICE NOTES

FM mobile.

**Frequency range**

430-440 MHz

**Output power**

25 W / 5 W

## IC-48 E

NOVICE NOTES

A replacement to the IC47E which had the facilities for Digital Code Squelch. This was a system which used a form of selective calling so that the set would remain quiet until you were called by someone who only wanted to talk to you. DCS was a system used by Icom & Kenwood but never became popular.

**Frequency range**
430-440 MHz
**Output power**
25 /5W

## IC -120

FM mobile.

**Frequency range**
1240-1300 Mhz
**Output power**
1Watt (or slightly greater)

## IC-449 E

Synthesised FM mobile transceiver.

**Frequency range**
430-440MHz
**Output power**
35W / 5W

## IC-μ4E

NOVICE NOTES

FM handheld transceiver. Introduced as a "pocket sized" handheld and is still one of the more compact sets today. The set is considered very rugged as it has a steel frame with high impact plastic mouldings for the main body. The USA version had an inbuilt CTCSS encoder with the option of CTCSS decode being added to allow total tone squelch operation. Check if the UK version had the CTCSS encode facilities built in as standard.

## IC-2E/2SE/2PE

First released 1980(2E), 2 meter FM handie. New version (2SE) released 1989

**Frequency range**

1260 - 1299.99MHz

**Output**

1.5 watts.

## IC-12E

1200 MHz FM handie transceiver

**Frequency range**

1260 - 1299.99MHz

**Output**

1.5 watts.

## IC-4SE

70cms FM Handie transceiver

**Frequency range**

430 -440MHz

**Output**

5 watts.

## IC32E

Dual band handie transceiver

**Frequency range**

144 - 146MHz and 430 - 440MHz

**Output**

5.5 W (VHF) 5.0W (UHF)

## IC24ET

Dual band handie transceiver

**Frequency range**

144 - 146MHz and 430 - 440MHz

**Output**

5,3,1.5 watts.

## ICW2E

Dual band FM handie transceiver

**Frequency range**

144 - 146MHz and 430 - 440MHz

**Output**

5.5 W (VHF) 5.0W (UHF)

## IC-W21E

Dual band FM handie transceiver

**Frequency range**

144 - 146MHz and 430 - 440MHz

**Output**

5.5 W (VHF) 5.0W (UHF)

## IC-505

6 meter multimode, mobile/base transceiver

**Frequency range**

50.000.00 - 53.999.9MHz

**Output**

10 watts with external supply, 3 watts with internal supply

**Modes**

SSB, CW, FM (Option)

## IC575

10 meter and 6 meter base transceiver

**Frequency range**

Receive
26.0000 - 56.0000MHz
Transmit
28.0000 - 29.7000MHz
50.0000 - 54.0000MHz

**Output**

10 watts with external supply, 3 watts with internal supply

**Modes**

SSB, AM, FM

## IC3200

25 watt mobile FM dual bander

**Frequency range**

144 - 146MHz and 430 - 440MHz

**Output**

25W, 5W

## IC-2SRE

2 meter and 70cms FM handie with wideband receive

**Frequency range**

Receive
25 - 950MHz
Transmit
144-146MHz
430-440MHz

**Output**

5W, 3.5W, 1.5W, 500mW

**Modes**

FM (Tx), Wide FM and AM receive

## IC-S21E/S41E

2 meter (S21E) or 70cms (S41E) compact handie transceiver

**Frequency range**

144 - 146MHz or 430 - 440MHz

**Output**

6 watts or several lower settings

## IC-2GE

2 meter handheld FM transceiver

**Frequency range**

144 - 146MHz

**Output**

7 watts

## IC-4E

70cms FM handie transceiver

**Frequency range**

430 -439.95MHz

**Output**

1.5W, 0.15W

## IC-3220E/H

Mobile dual band FM transceiver

**Frequency range**

144 - 146MHz and 430 - 440MHz

**Output**

25W (E model) 45 watts (H) model

## IC3230E/H

Compact mobile dual band FM transceiver

**Frequency range**

144 - 146MHz and 430 - 440MHz

**Output**

25W (E model) 45 watts (H) model

## IC-2400E

Mobile dual band FM transceiver

**Frequency range**

144 - 146MHz and 430 - 440MHz

**Output**

45W (VHF) 35 watts (UHF)

## IC-2500

Mobile dual band FM transceiver

**Frequency range**

430 -440MHz
1240 - 1300MHz

**Output**

35 watts, 10 watts

## IC-900

FM mobile dual transceiver

**Frequency range**

144 - 146MHz and 430 - 440MHz

**Output**

50W (VHF) 35 watts (UHF)

## IC-901

Mobile dual band FM transceiver

**Frequency range**

144 - 146MHz and 430 - 440MHz

**Output**

50W (VHF) 35 watts (UHF)

## IC-970

Multimode base station VHF and UHF transceiver. Large display

**Frequency range**

144 - 146MHz, 430 - 440MHz, 1240 - 1300 and 2400 - 2450MHz

**Modes**

SSB, CW, FM

**Output**

40 watts (FM), 30 watts (SSB and CW)

## IC-1271E

1.2GHz base station transceiver

**Frequency range**

1240 - 1300MHz

**Modes**

SSB, CW, FM

**Output**

10W, 1W

## IC-1201

Compact 1.2GHz mobile transceiver

**Frequency range**

1240 - 1300MHz

**Output**

10 watts, 1 watts

## IC-1200

Compact 1.2GHz mobile transceiver

**Frequency range**

1240 - 1300MHz

**Output**

10 watts, 1 watt

## IC-120

Compact 1.2GHz mobile transceiver

**Frequency range**

1260 - 1300MHz

**Output**

1 watt

## IC-471H

Base station UHF multimode transceiver

**Frequency range**

430 - 450MHz

**Modes**

SSB, CW, FM

**Output**

75 watts

## IC-451

Base station UHF multimode transceiver

**Frequency range**

430 - 450MHz

**Modes**

SSB, CW, FM

**Output**

25 watts

## IC-490
Multimode UHF base/ mobile station transceiver

**Frequency range**
430 - 450MHz

**Modes**
SSB, CW, FM

**Output**
10 watts

## IC-449
Mobile UHF FM transceiver

**Frequency range**
430 - 450MHz

**Output**
35W, 20W, 10W, 5W

## IC-448
Mobile UHF FM transceiver

**Frequency range**
430 - 450MHz

**Output**
35W, 5W

## IC-48
Mobile UHF FM transceiver

**Frequency range**
430 - 450MHz

**Output**
35W, 5W

## IC-47
Mobile UHF FM transceiver

**Frequency range**
430 - 450MHz

**Output**
25W, 5W

## IC-45
Mobile UHF FM transceiver

**Frequency range**
430 - 450MHz

**Output**
10W

## IC-251
2 meter base station multimode transceiver

**Frequency range**
144 - 146MHz

**Modes**
SSB, CW, FM

**Output**
10 watts

## IC-290
Released 1985 and retailing at £469. 2 meter multimode base transceiver

**Frequency range**
144 - 146MHz

**Modes**
SSB, CW, FM

**Output**
10 watts

## IC-229
Mobile 2 meter FM transceiver

**Frequency range**
144 - 146MHz

**Output**
50 watts

## IC-228
Mobile 2 meter FM transceiver

**Frequency range**
144 - 146MHz

**Output**
45 watts

## IC-28/IC-28H
Mobile 2 meter FM transceiver

**Frequency range**
144 - 146MHz

**Output**
25 watts, 45 watts (IC-28H)

## IC-27
Compact 2 meter mobile transceiver

**Frequency range**
140 - 149.995MHz

**Modes**
FM

**Output**
25 watts high, 5 watts low

# TS-950SDX

Base station HF transceiver with Digital Signal Processing on transmit and receive. Follow on from the TS-950 and TS-940, but with some novel new features. Extensive menu system. MOSFET power output transistors giving 150 watts output. Dual receive on same band with separate tuning.

**Frequency range**

| | |
|---|---|
| Receive: | 100kHZ - 30MHz |
| Transmit: | All HF bands between 1.8 and 29.70MHz |

**Modes:**

SSB, CW, AM, FM, DATA (RTTY)

**Output power:**

5-150 watts (SSB, CW, RTTY, FM), 5-40 watts (AM)

**Features:**

Power supply included
Slope pass band tuning
CW Reverse mode
Manual notch
1 Hz tuning
Two in band, simultaneous receivers - separate tuning
RIT
Digital filters can be on or off

# TS-870

First shown at Stafford during October 1995. Entirely digital (no mechanical filters) and DSP functions carried out at IF stages. See Peter Hart review in this Guide

**Frequency range**

Receive: 100kHZ - 30 MHz

Transmit: All HF bands between 1.8 and 29.70MHz plus 6 meters

**Modes:**

SSB, CW, AM, FM, RTTY

**Output power:**

5-100 watts (SSB, CW, RTTY, FM), 5-50 watts (AM)

**Features:**

Slope pass band tuning
CW Reverse mode
Manual notch
Auto notch (DSP)
1 Hz tuning
Two in band simultaneous receivers with separate tuning

# TS-570D

First shown at the RSGB 1996 HF Convention. Mid priced base station with Digital Signal Processing as standard. A new feature, CW auto tune is included which retunes the VFO frequency to the set pitch. Interestingly, the interference reduction facilities are described as being DSP at AF stages. This appears to run counter to the claims for IF stage processing on the TS-870.

**Frequency range**

| | |
|---|---|
| Receive: | 30kHZ - 30MHz |
| Transmit: | All HF bands between 1.8 and 29.70MHz |

**Modes:**

SSB, CW (AM and FM if options fitted)

**Output power**

10-100 watts SSB, CW (FM if fitted), 10-40 watts (AM if fitted)

**Features:**

CW auto tune
DSP AF noise reduction
DSP transmit audio shaping
Extensive memory features
Scanning
Personal computer RS232 interface via 9 pin D-Sub connection
VOX

# TS-50S

Compact mobile HF transceiver. Provoked a lot of interest when first released due to its small physical size and features. Forerunner to many similar compact transceivers released by other manufacturers.

**Frequency range**

| | |
|---|---|
| Receive: | 30kHZ - 30MHz |
| Transmit: | All HF bands between 1.8 and 29.70MHz |

**Modes:**

SSB, CW (AM and FM if options fitted)

**Output power:**

10-100 watts SSB, CW (FM if fitted), 10-40 watts (AM if fitted)

**Features:**

Optional FM, AM board
Dual VFO's with split capability
Band stacking
30 memory channels
10Hz resolution

KENWOOD

## TS-950/TS950SD

Top of the line, base HF station transceiver with general coverage receiver. The TS-950SD (released circa 1989) was identical to the
TS-950 but with all options fitted

**Frequency range**

Receive
150kHz - 30MHz
Transmit
All HF bands from 1.8 to 29.70MHz

**Modes**

SSB, CW, AM, FM, RTTY (FSK)

**Features**

TS-950SD:
Internal ATU
All filters fitted
DSP-10 unit
High stability crystal
Dual VFO
Notch filter
Slope tuning
Noise blanker
Variable IF passband width
Built in electronic keyer
100 memories
RF speech processor
AC power supply built-in

## TS-940S

General coverage, all HF bands transceiver. analogue S-meter, and internal ATU. The first Kenwood transceiver with "slope" display.

**Frequency range**

Receive
150kHz - 30MHz
Transmit
All HF bands from 1.8 to 29.70MHz

**Modes**

SSB, CW, AM, RTTY (FSK)

**Features**

Internal ATU
Analogue S-meter
Notch filter
CW variable bandwidth tuning
Audio filter
RIT
Fluorescent and LCD display
RF speech processor
AC power supply built-in

## TS-930S

General coverage, all HF bands transceiver. All solid state. Reputed to have an excellent receiver

**Frequency range**

Receive
150kHz - 30MHz
Transmit
All HF bands from 1.8 to 29.70MHz

**Modes**

SSB, CW, AM, RTTY (FSK)

**Features**

Internal ATU
Dual VFO
RIT
Notch filter
Noise blanker
RF speech processor
AC power supply built-in

HF RANGE

KENWOOD

HF RANGE

## TS-850S

Highly popular HF base station transceiver with general coverage receiver. Optional Digital Signal Processing module and internal antenna tuner.

**Frequency range**

Receive:
100KHz - 30MHz
Transmit:
All HF bands between 1.8 and 29.70MHz

**Modes:**

SSB, CW, AM, RTTY, FM and AM

**Output power:**

100 watts, 40 watts AM

**Features:**

1Hz tuning
DDS tuning system
Slope tuning
RF speech processor
VOX
Dual VFO's with split capability
Tunable notch filter
100 memories
Scan functions
Dual mode noise blankers
Optional voice recording system
RIT/XIT incremental tuning
Transmit HF boost function

## TS-450/690S

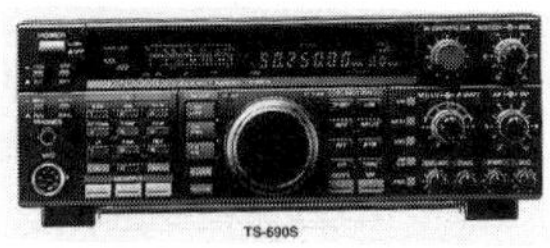

Compact and lightweight mid range HF base station transceiver with general coverage receiver. Optional Digital Signal Processing unit. Built-in antenna tuning unit option. The TS-690S has the addition of the 6 meter band.

**Frequency range**

Receive:
500kHZ - 30MHz(TS-450S) and
50-54MHz (TS-690S)
Transmit:
All HF bands between 1.8 and 29.70 and 6 meter band (TS-690S)

**Modes:**

SSB, CW, AM, FM, RTTY

**Output power:**

100 watts (SSB, CW, FM)
40 watts AM

**Features:**

100 memories
Split frequency
Twin VFO's
Notch filter

## TS-140/680S

Entry level HF base station transceiver with general coverage receiver.
TS-680S includes 6 meters

**Frequency range**

Receive:
500kHz - 30MHz (TS-140S)
and
50-54MHz (TS-680S)
Transmit:
All HF bands between 1.8 and 29.70 and 50MHz (TS680S)

**Modes:**

SSB, CW, AM, FM

**Output power:**

100 watts (SSB, CW)
50 watts (FM)
40 watts AM

**Features:**

31 memory channels
Dual noise blanker
Scanning functions
RIT (receiver incremental tuning)
Speech processor
10Hz tuning
Optional antenna tuning unit

KENWOOD

VHF/UHF RANGE

## TR-3200 NOVICE NOTES

Crystal controlled FM transceiver. Possibly limited by the fact that the set is crystal controlled and replacement crystals would now be needed to be ordered from specialist firms. It is larger than many of the more modern synthesised transceivers but could still be considered as an entry level set if fully fitted with crystals for useful channels. Otherwise consider a synthesised set. The set is not designed to cover the whole band and the transmit is limited to between 432 & 436MHz. Outside this range the power will drop.

**Frequency range**
432-436MHz
**Output power**
2W/ 400W

## TR9500

Multimode Mobile Transciever.

**Frequency range**
430-440 MHz
**Output power**
10 W / 1W

## TR8400 NOVICE NOTES

Synthesised FM Mobile. Early synthesised FM tranceiver, but for it's time quite advanced in that it offered 2 VFO'S & memory facilities.

**Frequency range**
430-440 MHz
**Output power**
10 W / 1W

## TR8300 NOVICE NOTES

22 channel crystal controlled FM Mobile. The set is limited by the fact that is crystal controlled and replacement crystals would now be needed to be ordered from specialist firms. The Transmit is limited between 432 & 438 MHz, outside this range the power will drop. Possibly worth considering if fully fitted with crystals. y The cost of adding additional channels would make a slightly more expensive synthesised set a far better prospect.

**Frequency range**
432-438 MHz
**Output power**
10W / 1W

## TR-50 NOVICE NOTES

FM transceiver. 23 cm's transceiver which was advertised briefly in the American market. Very rarely seen in the UK.

**Frequency range**
1240-1300 MHz
**Output power**
1W

## TM 421E NOVICE NOTES

FM mobile transceiver. Compact Mobile set with 14 memories available.

**Frequency range**
430-440 MHz
**Output power**
35W /5W

## TR-3500 NOVICE NOTES

FM handheld transceiver. The set uses an LCD frequency display. It is an early type of LCD display and may have suffered from some fading. The sets memories are backed up by a Lithium cell which has an estimated life of 5 years (manufacturer estimate). Check the set for full functionality before purchasing.

**Frequency range**
430-440 MHz
**Output power**
2W / 200mW

## TH41E

NOVICE NOTES

FM handheld transceiver. Very simple handheld, designed for basic functionality, possibly considered as the Trio-Kenwood answer to the IC-4E.

**Frequency range**
430-440 MHz

**Output power**
1W / 150 mW

## TR-3600E

NOVICE NOTES

FM handheld transceiver. This set was the successor to the TR-3500 E and included DCS (Digital Code Squelch, a Kenwood proprietary Selcall System. Improvements were made to scanning as well as storage of repeater splits in the memory channels.

**Frequency range**
430-440 MHz

**Output power**
2W / 200mW

## TM401A

NOVICE NOTES

FM mobile transceiver. A very compact set which was designed to fit in modern cars with very little room for equipment. This was acheived by making the speaker a separate part of the set rather than building it into the set.

**Frequency range**
430-440 MHz

**Output power**
10W / 1W

## TM411E

NOVICE NOTES

FM mobile transciever. This was based on the TM401A (see separate entry) and differed in that it included the Trio-Kenwood Digital code squelch system, which allowed paging and Auto QSY between suitably equipped sets. The set also has a tiltable front panel so that the display can be positioned for ease of reading in mobile situations.

**Frequency range**
430-440 MHz

**Output power**
25W / 5W

## TR9300

NOVICE NOTES

Multimode mobile transceiver. A 6 meter mobile transceiver, with twin VFO'S and 6 memories.

**Frequency range**
50-54 MHz.

**Output power**
25W / 5W

## TS-811E

NOVICE NOTES

Multimode base station. Launched as a de-luxe base station set, it was in production for many years before being deleted. The set came equipped with Digital Code Squelch which is a proprietary Selcall System. The set announces changes of mode with morse letters, for example, F for FM & U for USB while the optional VS1 allows full voice announcement of frequency and any repeater shift. There is also the facility for automatic mode selection to match the band plan.

**Frequency range**
430-440 MHz

**Output power**
25W Variable

## TM-431E

NOVICE NOTES

FM mobile transceiver. The set was part of a family covering 2cm, 70cm, 23cm & a 2cm / 70cm dual band set. The RC-20 controller with the IF-20 could control up to 4 sets. The microphone, as well as having the usual Up/Down buttons had other control facilities to make the set easier to use when mobile, and allowed you to set the memories, VFO'S & CTCSS frequencies. Another button can be set by the operator.

**Frequency range**
430-440 MHz
**Output power**
10W/ 5W

## TM-531E

NOVICE NOTES

FM mobile transceiver. The set was part of a family covering 2cm, 70cm, 23cm & a 2cm / 70cm dual band set. The RC-20 controller with the IF-20 could control up to 4 sets. The microphone, as well as having the usual Up/Down buttons had other control facilities to make the set easier to use when mobile, and allowed you to set the memories, VFO'S & CTCSS frequencies. Another button can be set by the operator.

**Frequency range**
1200-1300MHz
**Output power**
10W / 5W

## TS-60S

NOVICE NOTES

Multimode mobile. This is the 6 meter only version of the popular TS-50S HF mobile transceiver. Although the power output is way above that permitted under the terms and conditions for the Novice Licence, it is impossible to reduce the output power. It features some innovative ideas as regards the internal control of the set by the microprocessor.

**Frequency range**
50-54 MHz
**Output power**
90W / 50W / 10W

## TH-48E

NOVICE NOTES

FM Transceiver. The set is a single band transciever with dual band receive capabilities. The transceiver also comes with DTMF capabilities inbuilt and allowes alpha numeric paging.

**Frequency range**
430-440 MHz
**Output power**
between 200mW & 5W depending on supply

## C156E

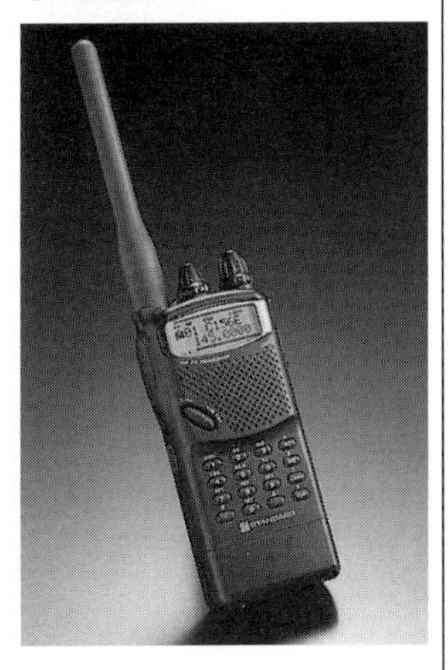

VHF FM handheld transceiver. Available in the UK from October 1996.

**Frequency range**

Receive:
144 - 147.995
Transmit:
144 - 147.995

**Modes:**

FM

**Output power:**

3W, 2.5W, 0.35W

**Features:**

7 methods & 3 types of scan
8 tuning steps 5/10/12/15/20/25/30/50 kHz
39 tone encoder sequences built-in
1750 Hz tone burst built-in

## C568

VHF and UHF (triple band) handheld transceiver with 23cm, 70cm and 2 meters

**Frequency range**

144-147.995, 430-433.995, 1240-1299.990MHz
Airband (receive)

**Modes:**

FM

**Output power:**

5.0W or 2.5W depending on nicad in use

**Features:**

Very compact
Twin band receive
Full duplex
40 channels
8 tuning steps
Twin band dual watch
Wake up feature
Tone burst built-in

## C116/C416

Dual band handheld transceivers.

**Frequency range**

144-147.995 with sub-band 430-439.995 (C116)
430-439.995 with sub-band 144-147.995 (C416)
Airband receive

**Modes:**

FM

**Output power:**

5W or 1W depending on nicad in use
20mW for sub band

**Features:**

Semi duplex operation
8 scan modes
7 tuning steps
100 memories
Tone burst
Wake up feature
Dual watch

## C108S/C408

Ultra compact single band FM (shirt pocket size) transceiver. C108 is a VHF version, C408 the UHF version.

**Frequency range**

144-145.995 (C108S)
430-439.995 (C408)

**Modes:**

FM

**Output power:**

230mW

**Features:**

Simple and basic
20 channels
41 hour battery life
Programmable scan
Only 130 g weight including batteries

## C 7800

NOVICE NOTES

FM mobile transceiver. An early synthesised UHF mobile but extremely easy to use. The set has 4 memory channels but these are cleared if all power is disconnected. The C7800 is extremely well built internally, with each section being screened. It has a very sensitive receiver. With the addition of a suitable PSU, it would make an ideal set for Novices to use at home or when mobile.

**Frequency range**

430 - 439.975 MHz (25 or 50 KHz steps)

**Output power**

10 W / 1W

## C 78

FM portable.

**Frequency range**

430-440 Mhz

**Output power**

1Watt

## C7900

NOVICE NOTES

FM mobile transceiver. The replacement set for the C7800. Very rarely seen. I can only remember seeing one advertised recently. The set is in a very slim line case which would make it ideal for locating in cars with only a limited amount of room.

**Frequency range**

430-440 Mhz

**Output power**

10W / 1W

# FT-1000

Flagship HF bands base station transceiver with 200 watts output, dual receiver (cross band with bandpass option). Large display and two flywheel-weighted tuning knobs.

**Frequency range**

Receive: 100kHZ - 30MHz

Transmit: All HF bands between 1.8 and 29.70MHz

**Modes:**

SSB, CW, AM, FM

**Output power:**

200 watts

**Features:**

Built-in Antenna Tuner
Power supply built-in (AC version)
100 memory channels
DDS tuning system
200 watts output
Optional digital voice recorder
RF speech processor
Dual receivers with separate tuning (different bands with band-pass option)

# FT-1000MP

First shown at Stafford 1995. New and popular addition to the Yaesu range and despite its name, not a replacement for the FT-1000. Many of the design features are a result of the research and development carried out for the FT-1000. Digital signal processing is included for transmit and receive. Dual receivers are included with separate tuning knobs. Collins mechanical filters are available as an option. Two supply versions available, AC included or DC only.



**Frequency range**

| | |
|---|---|
| Receive: | 100kHZ - 30MHz |
| Transmit: | All HF bands between 1.8 and 29.70MHz |

**Modes:**

SSB, CW, AM, FM, RTTY

**Output power:**

Fully adjustable 5-100 watts (SSB, CW, RTTY, FM), 5-25 watts (AM)

**Features:**

Digital Signal Processing at receive and transmit
Clarifier tuning
Collins filters
Dual in band receive with separate S meters
Selectable antenna jacks
User programmable tuning steps
Extensive menu system allows customising
Built-in antenna tuner

# FT-990

HF base station transceiver with a general coverage receiver. Comes in two versions, one with AC power supply included or DC only, where a separate supply is required.

**Frequency range**

| | |
|---|---|
| Receive: | 100kHZ - 30MHz |
| Transmit: | All HF bands between 160-10m |

**Modes:**

SSB, CW, AM, FM, RTTY

**Output power:**

100 watts SSB

**Features:**

Built-in antenna tuner
RF speech processor
Dual VFO's
90 memories
Band stacking system
Optional digital voice recorder
Separate antenna sockets selectable from front panel

# FT-900AT

HF base or mobile transceiver, being one of the first designs with a detachable front panel for mobile mounting. Built-in antenna tuner makes it an ideal HF mobile rig.

**Frequency range**

| | |
|---|---|
| Receive: | 100kHZ - 30MHz |
| Transmit: | All HF bands between 1.8 and 29.70MHz |

**Modes:**

SSB, CW, AM, FM, RTTY

**Output power:**

5-100 watts (SSB, CW, FM) 25 watts on AM

**Features:**

Keypad entry when used as base station rig
IF shift
Notch filter
Programmable CTCSS encode with repeater offset
DDS tuning system
100 memories
Twin band stacking VFO's
Built-in noise blanker
Speech processor
Collins filters

# FT-840

Compact budget priced mobile or base station HF transceiver. Dual superheterodyne receiver with DDS tuning system.

**Frequency range**

Receive: 100kHZ - 30MHz

Transmit: All HF bands from 1.8-29.7MHz

**Modes:**

SSB, CW, AM, FM (with option fitted)

**Output power:**

Variable to 100 watts (AM 40 watts)

**Features:**

1 Hz tuning
Optional antenna tuner
100 memories
Built in electronic keyer
IF shift
Noise blanker
AF speech processor
Optional filters
FM option available
Twin VFO's

# FT-736R

Multimode VHF/UHF base station transceiver. Especially suitable for Satellite operation. With options can cover up to 4 of the 50, 144, 220, 430MHz and 1.2GHz bands. Supplied with built-in power supply or can be powered by a separate power supply.

**Frequency range**

| | |
|---|---|
| Receive: | 144-146MHz and 430-450MHz<br>can be extended with options for 50MHz, 220MHz and 1.2GHz bands |
| Transmit: | 144-146MHz and 430-450MHz<br>can be extended with options for 50MHz, 220MHz and 1.2GHz bands |

**Modes:**

USB,LSB,CW AM,FM

**Output power:**

25 watts variable, 10 watts on 50 and 1240MHz

**Features:**

Power supply built-in
100 memory channels
Full cross band duplex with inverted tracking for A0-13
TNC interface
Relay control for linear amplifier
Optional band modules

## FT-8500

Dual band FM mobile transceiver

**Frequency range**

Receive
110-180MHz (VHF)
420-470MHz (UHF)
Transmit
144-146MHz
430-450MHz

**Modes:**

FM, AM Aircraft receive

**Output power:**

50W, 10W, 5W (VHF)
35W, 10W, 5W (UHF)

**Features:**

110 memories
Packet connector
Auto power off
Cross band
5 Scanning modes

## FT-8000R

Dual band FM mobile transceiver

**Frequency range**

Receive
110-550MHz(VHF)
750-1300MHz (UHF)
Transmit
144-146MHz
430-450MHz

**Modes:**

FM

**Output power:**

50W, 10W, 5W (VHF)
35W, 10W, 5W (UHF)

**Features:**

110 (55 per band) memories
1200/9600 Packet compatible
Auto power off
Cross band repeat
Scanning facilities
Dual receive
Detachable remote front panel
Choice of microphones
Wide receiver coverage

## FT-5100

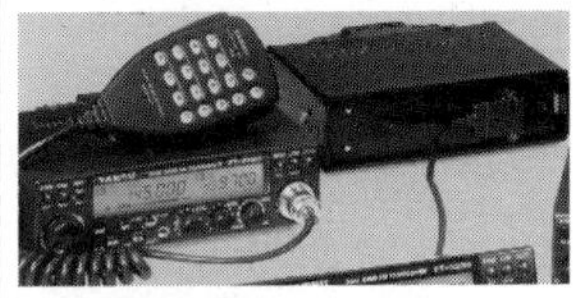

Dual band FM mobile transceiver.

**Frequency range**

130-174MHz
430-450MHz

**Modes:**

FM

**Output power:**

50W, 5W (VHF)
35W, 5W (UHF)

**Features:**

94 memories
Dual in-band receive
Full duplex cross band
Built in antenna duplexer
CTCSS encode built-in

## FT-3000M

High powered (70 watts) 2 meter FM mobile transceiver. Wide band receive, VHF and UHF up to 999MHz. AM aircraft receive.

**Frequency range**

Receive
110-180MHz
300-520MHz
800-999MHz
AM aircraft
Transmit
144-146MHz

**Modes:**

FM

**Output power:**

70W, 50W, 25W, 10W

**Features:**

81 memory channels
1200/9600 packet compatible
Smart search
Digital coded squelch
Scrolling menu system
MIL-STD-810 rating

## FT-2500M/7400H

Popular dual-band (VHF - UHF) handie. Very small size. Released 1996.

**Frequency range**

Receive
140 - 174MHz FT2500M
430-450MHz FT7400H
Transmit
144- 146MHz (FT2500M)
430-450MHz (FT7400H)

**Modes:**

FM

**Output power:**

50, 20, 5W
35, 15, 5

**Features:**

MIL-STD 810 rating
31 memories
CTCSS encode built-in
5 scanning functions

## FT-2200/7200

2 meter (FT2200) or 70cm (FT-7200) FM mobile transceiver

**Frequency range**

Receive
110 - 180MHz (FT2200)
430-450MHz (FT7200)
AM aircraft receive
Transmit
144- 146MHz (FT2200)
430-450MHz (FT7200)

**Modes:**

FM

**Output power:**

50, 20, 5W (FT2200)
35, 15, 5 (FT7200)

**Features:**

Compact mobile
50 memories
AM Aircraft receive

## FT-530

Dual band handheld FM transceiver

**Frequency range**

Receive:
130 - 174MHz
430-450MHz
Transmit:
144-146MHz
430 - 450MHz

**Modes:**

FM

**Output power:**

5, 2, 1.5, 1 or 0.5 watts

**Features:**

82 memories
Dual in-band receive
CTCSS encode built-in
Built-in VOX

## FT-290/690/790RII

Field operation mobile mutimode transceivers

**Frequency range**

144-146MHz (FT290RII)
50-54MHz (FT690RII)
430-450MHz (FT790RII)

**Modes:**

FM, SSB, CW

**Output power:**

25W FT290RII
10W FT690RII
25W FT790RII

## FT-51R

Dual band handheld FM transceiver

**Frequency range**

Receive:
130-174MHz
430 -450MHz
AM Aircraft
Transmit:
144 - 146MHz
430 - 440MHz

**Modes:**

FM

**Output power:**

5 selectable levels

**Features:**

120 memories
Scrolling user help menu
Automatic tone search
3 dual receive configurations
DTMF paging built-in
3 scan modes

## FT11/41R

YAESU

VHF/UHF RANGE

Mono band handheld FM transceivers

**Frequency range**

Receive:
110 - 180MHz (FT11R)
420-470MHz (FT41R)
AM Aircraft
Transmit:
144 - 146MHz (FT11R)
430 - 440MHz (FT41R)

**Modes:**

FM

**Output power:**

5, 1.5 watts

**Features:**

150 memories

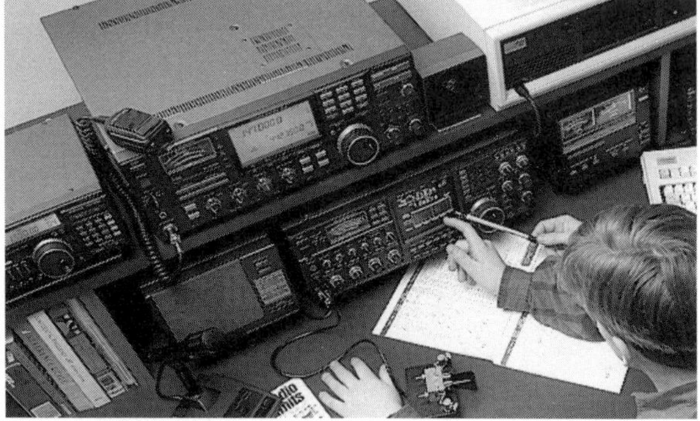

*Amateur Radio is a challenging, exciting and fun hobby for all ages*

## FT690 R

50 MHz multimode.
Replaced by FT690 R MK2

**Frequency range**
50-54 MHz
**Output power**
2.5 Watts

## FT690 R MK2

NOVICE NOTES

6 meter multimode transceiver. The Mark 2 was introduced to replace the early Mark 1 series. The front panel was completely redesigned to improve the ergonomics, though some people are known to have said that the Mark 1's were the better looking set. A problem with the batteries was resolved with Mark 2. The amplifier for the MK 2 clips in place of the battery pack and provides power for the set as well as the amplifier.

**Frequency range**
50-54 MHz
**Output power**
2.5 Watts

## FT790 R

70cm multimode.
Replaced by FT790 R MK2.

**Frequency range**
430-440 MHz
**Output power**
1Watt / 0.1Watt

## FT790R MK2

NOVICE NOTES

70cm multimode transceiver. The amplifier for the Mark 2 clips in place of the battery pack and provides power for the set as well as the amplifier.

**Frequency range**
430-440 MHz
**Output power**
1Watt

## FT708R

NOVICE NOTES

Keyboard controlled FM Handheld. The set has a LCD display with backlight for frequency indication. Battery back up of memory and shifts etc. The battery life was stated as 5 years. Check that the battery is still ok, and also, as the LCD is of an early type, that the LCD is still readable.

**Frequency range**
430-440 MHz 25 kHz steps.
**Output power**
1W or 0.1W

## FT-780 R

NOVICE NOTES

Mobile multimode. There were two versions of this set available. The early versions had a 7.6 Mhz repeater shift and the FT780R had a 1.6 MHz UK repeater shift fitted (this cost £10 more). The early versions could still be used with UK repeaters by operating with 2 VFO'S, VFO A for rx & VFO B for tx. There were also some versions imported under the Sommerkamp Brand name.

**Frequency range**
430-440 MHz 2 VFO's (steps according to mode)
**Output power**
10W / 1W

## FT-680 R

NOVICE NOTES

Mobile multimode. This is the 6 meter version of the FT-780, and to the best of my knowledge there were very few of these sets sold in the UK, as just after the band was opened for general release, the set was discontinued. There were also sets to the same specification under the Sommerkamp brand name.

**Frequency range**
50-54 MHz 2 VFO's (steps according to mode)
**Output power**
10W / 1W

## FT-404R

NOVICE NOTES

Crystal controlled FM transmitter. The set is very limited these days, being crystal controlled. The cost of obtaining extra crystals for just the different repeaters around the country should be considered as against purchasing a slightly more expensive synthesised set.

**Frequency range**
430-440 MHz

**Output power**
2.5W

## FT-6200

NOVICE NOTES

Dual band FM mobile. The set is a dual band 70 & 23 cms transceiver which allows the user to operate full duplex (transmitting on one band while receiving on the other). There is the option of remote mounting the main body of the transceiver whilst only having the front panel on display. The set comes aquipped with full CTCSS encode on both bands, the decoding is an optional extra.

**Frequency range**
430-440 MHz & 1240 - 1300

# The IC-775DSP and FT-1000MP Compared

***by Peter Hart, G3SJX***

WHEN IT COMES to HF transceivers, every once in a while a new innovation arrives on the scene which sets the trend for future designs. The last major innovation, several years ago, was the direct digital synthesiser which revolutionised the design of the frequency synthesiser. We are now on the verge of another significant breakthrough, the advent of built-in digital signal processing (DSP) to give a level of signal processing flexibility unattainable by more conventional analogue techniques. DSP is by no means new and has been used in commercial and military HF radios for quite some time. In the amateur market, the use of DSP has been confined to add-on boxes giving sophisticated levels of audio filtering, tracking automatic notch filters, noise reduction and similar low frequency processing tasks.

We are now seeing the domain of DSP technology moving forwards in the architecture of the radio, encompassing the demodulation and IF functions. This promises to offer a degree of performance and features as yet unseen in HF transceivers. During the last few months, all three of the big Japanese suppliers, Yaesu, Icom and Kenwood have launched new top of the range models. These radios all employ DSP to a greater or lesser extent in the backend signal processing but, much more than this, a level of features is provided which is more comprehensive than any other radios produced to date. This review looks at the Icom IC-775DSP and Yaesu FT-1000MP models. The Kenwood TS-870 will be reviewed in the very near future.

The new Icom flagship is the IC-775DSP, a new design. The FT-1000MP is a major update of the existing FT-1000, offering many more features and improvements and at a lower cost. The transmit power output has been reduced from 200W to 100W, which is probably the major cost saving, and both versions will continue to be marketed. Both the IC-775DSP and FT-1000MP are substantial mains powered base station radios and the FT-1000MP can in addition be powered from 12V DC. The Icom radio measures 424W x 150H x 390D mm and weighs 16.7kg. The Yaesu radio is slightly smaller at 410W x 135H x 347D mm and weighs 15kg. Neither radio is fitted with carrying handles.

## IC-775DSP PRINCIPAL FEATURES

TWIN RECEIVERS ARE provided and the specified tuning range of both is from 100kHz to 30MHz, although the tuning range actually extends down to 30kHz. On transmit, the frequency range covers the widest amateur allocations used in the world, although there are national variants available. Modes provided are USB, LSB, CW, FM, AM and RTTY. Data is supported in AFSK or FSK modes with several selectable tone frequencies and spacings in RTTY FSK mode. Data mode may be selected on all voice modes. This mutes the microphone and hence enables AFSK (packet / SSTV / RTTY etc) to be used on any mode without unplugging the microphone. The display indicates the carrier frequency on data modes. Both USB and LSB may be selected on CW (normal / reverse CW) which can be useful for combating QRM.

Two rotary tuning knobs are provided, one large and one small, for the main and sub receivers respectively. Three tuning step sizes are selectable - fine at 1Hz per step, 500 steps per knob revolution; normal at 10 / 20 / 50Hz per step, 500 steps per knob revolution; and quick at 1 - 10kHz per step, 250 steps per knob revolution. These step sizes are also mirrored in the sub receiver tuning although the steps per revolution are about half. Auto speed-up is engaged when the knob is turned rapidly. Individual buttons select bands and triple band stacking registers are provided. If a band key is pushed once, the last used frequency and mode on that band are recalled. A second or third set of frequency and mode are selected by subsequent key presses. This is very useful when up to three modes are frequently used. The main and sub receiver frequencies may both be input directly from the numeric keypad and UP / DOWN keys

will also step the frequency in programmable intervals from 1kHz to 1MHz. This is most conveniently left at 1MHz to provide band setting in general coverage mode or at say 20kHz for rapid steering within the amateur bands.
There are 101 memory channels provided, of which 99 will store one frequency and one mode. The other two are intended to store frequency limits for scanning. There is no provision to partition the memory as is done on many radios, and stepping through the whole bank is a fairly slow process. There are the usual read / write operations, including direct tune from any memory location to any frequency. Separate memory channel select and access keys are provided for the main and sub receivers, and the sub receiver can be used for memory preview. In common with other recent Icom radios, a memopad feature is provided. This is a quick and easy one touch memory store and recall facility which stores up to 10 frequencies on a stack. This is read out sequentially, last in, first out.
Twin VFOs are provided, with their separate tuning knobs to provide the dual receiver capability and split frequency transmit operation. These may be equalised and swapped and a single key allows reception on the transmit frequency in single receiver mode. Two further enhancements have been added to speed up split frequency operation - quick split and channelised split. Quick split enables the transmitter to be set to a pre-programmed offset (eg up 10kHz) by a single key press. Channelised split enables the transmitter to be stepped in frequency in pre-programmed steps.
Clarifier or independent tuning of the receiver (RIT) and / or of the transmitter are provided over a range of +/- 9.99kHz in 1Hz or 10Hz steps and the offset is displayed to 10Hz resolution. The offset may be cleared or added to the VFO frequency. Three scan modes are provided, scanning between two frequency limits, scanning of all occupied memory channels and scanning of selected memory channels only. Scan resume condition and scan speed are selectable.
Wide, medium and narrow IF bandwidth settings are selectable at both the second (9MHz) and third (455kHz) IFs for each of the various modes. The radio is fitted with 500Hz bandwidth CW filters in both these IFs as standard but other wider and narrower filters, including 250Hz CW, are optional extras. Twin passband tuning controls are provided for independent tailoring of the second and third IF passbands, and an IF notch filter at the 4th (10.7MHz) IF. A variable frequency audio peaking filter is fitted for CW use and various DSP filtering functions.
The receiver front-end configuration may be optimised to suit differing requirements. One of two switchable preamplifiers may be selected and three levels of signal attenuation. Other receiver functions include fully adjustable noise blanker, adjustable AGC speed and all-mode squelch / CW pitch control. The CW pitch may be set over the range 300 - 900Hz for most comfortable operation.
The transmit power is variable up to 200W output and transmit functions include RF based speech compressor, built-in keyer, full and semi break-in on CW, VOX, sub-audible tone encoder for FM repeaters and audio monitor.
A conventional analogue meter is fitted which shows power output, SWR, ALC or compression levels. An auto-ATU is included for matching antennas up to about 3:1 VSWR. This functions on transmit only and is bypassed on receive, and can be set to switch on automatically when the antenna match is poor.
Band stores are provided to give fast tuning by returning to the last used settings when a band is selected. Two separate antennas may be connected and selected from the front panel. Band stores memorise which antenna is used on which band and select the appropriate antenna.
The display is clear and bright with two settings for illumination. The main frequency is normally displayed to 10Hz resolution but can be indicated to 1Hz resolution, with sub frequency and IRT to 10Hz resolution. The usual mode and status messages are displayed and both main and sub selected memory channel numbers.
DIN accessory sockets are provided on the rear panel to interface to external auto-ATUs, data terminals, Icom linear etc. These have been common with all Icom transceivers for many years now and ensures interchangeability with all Icom accessories. Relay controlled T/R switching and ALC is also provided for general linear use, and the standard Icom CI-V serial computer interface.
Two key jacks are provided, the jack on the front panel for

a keying paddle to control the internal electronic keyer, and the jack on the rear panel for a straight key or external keyer. Apart from the two selectable antenna connections, there is also provision for external receive antenna or use with external receiver. Unusual these days is a low level RF output for transverter use.

## IC-775DSP ADDITIONAL FEATURES

THE DSP UNIT built into the IC-775DSP introduces some exciting new features. Digital noise reduction enhances the signal to noise ratio on received SSB, RTTY and SSTV signals. An auto notch function effectively tracks and notches out multiple heterodynes on SSB signals. This cannot be used on CW, but the manual notch is effective on this mode and both manual and auto notches can be used together. Digital modulators and demodulators are implemented which claim to give improved results through superior carrier and unwanted sideband suppression. An ultra narrow 80Hz bandwidth audio filter may be selected on CW which automatically tracks the CW pitch setting. The low and high cut-off frequencies for both the receive and the transmit audio passband may be tailored to suit individual preferences.

The electronic keyer built into the IC-775DSP has a variable weight setting. Various parameters of the paddle are settable and it is even possible to programme the UP / DOWN keys on the microphone to function as a CW keying paddle! Three 40 character memory stores may be programmed with contest messages, for example, and sent either at the same speed as the main keyer or at a different speed, and a programmable auto-repeat may also be selected.

It is now commonplace for many of the radio functions to be customised by some form of menu selection. The IC-775DSP carries this further and takes eight pages of the manual to describe. 26 categories of functions may be customised, many of these selecting sub-menus for further customisation.

## FT-1000MP PRINCIPAL FEATURES

THE TWIN RECEIVERS in the FT-1000MP each tune from 100kHz to 30MHz and the transmitter covers 500kHz wide segments around the amateur bands. Modes covered are USB, LSB, CW, AM, FM, RTTY and PACKET with independent modes on the two receivers. Both USB and LSB may be selected on CW (normal / reverse CW), normal or synchronous tuning on AM, SSB or FM modes on packet (FSK or AFSK) and USB or LSB modes on RTTY.

There are many ways of tuning the FT-1000MP. Two rotary tuning knobs set the main and sub receivers, tuning in a selection of step sizes from 0.625Hz up to 20Hz per step on CW and SSB modes, with 1000 steps per revolution of the knob. Default settings are 10Hz per step / 10kHz per revolution. 100Hz per step / 100kHz per revolution is used on AM and FM and all step sizes and tuning rates are multiplied by 10 by pressing a FAST key. UP / DOWN keys step in 100kHz or 1MHz increments and a rotary click step control tunes in programmable steps of 1-100kHz (default 10kHz) for fast navigation. This click step control also doubles as the memory selector. Individual keys select the different bands and double band stacking registers are used. The bands may be set separately for the main and sub receivers. The frequency of the main and sub receivers may be entered directly from the numeric keypad.

A totally new method of tuning is implemented in the FT-1000MP, 'shuttle jog tuning'. This employs a spring loaded ring around the outside of the main tuning knob. Moving the ring away from its centre detent position in either direction, engages constant tuning, and the further the ring is moved the larger are the tuning steps. Thirteen preset steps from 10Hz to 100kHz are progressively selected as the ring is turned, enabling either small frequency changes or large frequency changes to be easily accomplished.

Ninety-nine regular memories are provided, nine programmed limit memories and five quick access memories. Each memory location stores dual frequencies (main and sub), modes, filter selections, clarifier settings and split status. The regular memories may be partitioned into groups, up to five, and memory selection is rapid with a click step rotary control. All the usual memory facilities are available and the sub receiver display is used to preview the memory contents whilst still retaining active use of both the main and sub receivers.

The quick memory bank facility functions in a similar way to the IC-775DSP. This uses store and recall buttons to store up to five frequencies

on a stack.
Split frequency facilities are similar to the IC-775DSP, including quick split but without channelised split. The clarifier functions on receive and / or transmit up to +/- 9.99kHz and on FM, a +/- 100kHz repeater offset is selectable. The repeater mode also transmits a selectable CTCSS sub-audible access tone. The full set of CTCSS tones is programmable.
Scanning is also provided covering band scan and various memory scan modes.
IF filter bandwidths of 250Hz, 500Hz, 2.0kHz, 2.4kHz are implemented at both the second and third IFs, with 6.0kHz at the third IF only. The 250Hz and 2.0kHz filters are optional extras and the others are standard fitment. An optional 500Hz filter is also available for the sub receiver IF. The review radio was fitted with all filter options. IF shift and width controls are fitted and a variable IF notch. Other filtering options are selected with the EDSP function.
Three alternative receiver RF amplifiers may be selected, a wideband amplifier covering the whole receiver range, a tuned low gain amplifier covering 1.8 - 7MHz and a high gain low noise tuned amplifier covering 24 - 30MHz. In addition the RF amplifier may be switched out (IPO) or three levels of attenuation inserted. Other receiver functions include two adjustable noise blankers, fast / slow AGC, separate squelch controls for main and sub and CW pitch over the range 300 - 1050Hz.
The transmit power is variable up to 100W output and transmit features are similar to the IC-775DSP, ie RF based speech processor, built-in keyer, full and semi break-in on CW, VOX, sub-audible CTCSS tone encoder, audio monitor and auto-ATU with band stores. Bargraph metering on the main display shows power output plus one of six selectable parameters. The multi-coloured fluorescent display shows more information than any other radio available. This includes main, sub and clarifier frequencies simultaneously to 10Hz resolution, twin bargraph S meters for main and sub receivers, separate twin bargraph displays for transmit functions, separate bargraph displays for tuning data signals and display offsets, memory channel groups and numbers and a host of status messages. All the bargraph displays have peak hold segments. The display is fairly bright but unlit segments are rather intrusive.
Two separate antennas may be connected and selected from the front panel and also separate receive antenna, or antenna to separate receiver. These antenna selections are stored with the band stores.
Rear panel interfacing includes DIN connectors for RTTY, packet, band data and DVS-2 voice store. The band data is used to interface to Yaesu linears and remote tuners. Connections are provided for audio in / out, linear switching and ALC, and low level RF for transverter drive. It is possible to display the transverted frequency on the FT-1000MP display when operating on 50, 144 or 432MHz. Two key jacks are provided, one on the front panel and the other on the rear and two headphone jacks, one quarter inch and the other 3.5mm.
The computer interface in the FT-1000MP may be directly connected to a PC serial interface with no external level converters. A nine pin D connector is used with standard pinning and a data rate of 4800 bits / sec. Commonly available computer cables can be used. Apart from full control of the front panel functions, there is also full computer access to the electronic keyer functions, EDSP, barmeters and menu programming settings.
Operation is comprehensively described in 11 pages of the operating manual. Note that the data reading format is different to the FT-1000, which means that existing FT-1000 software will need to be modified for use with the FT-1000MP. Turbolog 3 control and logging software has already been suitably modified (V3.07).
A very comprehensive and informative manual is provided with the FT-1000MP, containing 104 pages of information plus circuit diagrams.

## FT-1000MP ADDITIONAL FEATURES

YAESU DESCRIBES their DSP unit as EDSP - 'Enhanced Digital Signal Processor'. This has many functions similar to the Icom unit. Digital modulation and demodulation is provided and various filtering arrangements for the receive and transmit audio path. This provides compensation for various microphone or voice characteristics giving clearer audio. One of four different noise reduction algorithms may be implemented on receive to improve readability under various conditions and an auto notch will effectively track and notch out multiple heterodynes on SSB signals.

Four different filtering contours may be selected to improve readability under difficult conditions, providing bandpass, low, mid or high cutoff to the passband.

The electronic keyer built into the FT-1000MP has two iambic modes, bug key emulation and variable weighting capability. A fully featured contest memory keyer is also built-in which provides four 50-character messages, auto serial number incrementing and callsign ident. It is controlled via a remote keypad plugged into the rear panel which can also be used to provide certain memory and VFO control functions. No keypad is available yet from Yaesu, but it is simple to construct and full details are given in the manual.

The menu customisation provided by the FT-1000MP is perhaps even more extensive than the IC-775DSP. Some 79 of the transceiver settings are programmable, even down to providing fine trimming of the various receive and transmit carrier oscillator frequencies. Certain memory items tend to be accessed more frequently than others and these may be accessed via short cut keystrokes.

Several features have been incorporated to ease operation on digital modes - RTTY, AmTOR and packet. All the usual RTTY shifts and tone pairs are selectable on FSK with normal or reversed polarity. For 300 baud FSK packet, the IF passbands can be shifted to accommodate four different tone pairs and the frequency display offset to match. 1200 baud FM packet is also selectable and the microphone is muted on all digital modes. The frequency display can be set to indicate the mark tone or the suppressed carrier frequency on RTTY and the centre tone frequency or carrier on packet. A bargraph tuning meter is employed to indicate precise tuning on RTTY and packet signals. This is very effective and is also effective on CW.

## CIRCUIT DESCRIPTIONS

THE IC-775DSP RECEIVER is a quadruple conversion superhet on all modes except FM with IFs of 69.01MHz, 9.01MHz, 455kHz and 10.7MHz. FM mode only uses the first three IFs. The final IF is then converted to 15kHz to implement the DSP functions, so really it is a quintuple superhet receiver. The transmit signal is generated by the DSP at 15kHz and converted through 455kHz, 9MHz and 69MHz to final frequency.

The FT-1000MP main receiver is a quadruple conversion superhet with IFs of 70.455MHz, 8.215MHz, 455kHz and 10.24kHz. As with the IC-775DSP, FM uses only the first three IFs. The sub receiver is double conversion with IFs of 47.21MHz and 455kHz. EDSP is implemented at the 10.24kHz IF using an NEC uPD77016 16 bit DSP device. The transmit signal is generated by the DSP at 10.24kHz and converted through 455kHz, 8.215MHz and 70.455MHz to final frequency.

## SECOND RECEIVERS

THERE IS A MAJOR difference in the way that the second receiver function is implemented in these two transceivers. The IC-775DSP uses separate receiver first mixers which feed into a common signal path for all the IF, AF and DSP functions. The receiver input filters are also common to both front end mixers. Hence the Icom second receiver will use the same IF bandwidth and mode as the main receiver and must be broadly within the same or, in some cases, adjacent bands. Stereo copy, diversity etc are not possible.

The FT-1000MP uses common receiver input filters for both receivers but then totally separate signal paths through to the audio output. Although there is a similar restriction on input frequencies to the Icom, the main and sub receivers can operate with different IF bandwidths and different modes if needed. The main advantage is the ability to feed outputs to stereo headphones and copy both receivers in different ears. Mono, stereo and enhanced mixed modes are provided. It is also possible to lock the two receiver VFOs and with separate antennas on the main and sub receiver achieve full diversity reception. This opens up a wealth of possibilities for real marginal DX communication.

## MEASUREMENTS

THE MEASUREMENTS made on the radios are detailed in the table. Additional comments are as follows.

### *RECEIVER MEASUREMENTS*

The IC-775DSP sensitivity on 21 - 28MHz may be improved by 3dB over the 'Preamp In' figures by selecting the high gain preamp. The large change in FT-1000MP sensitivity and S9 level with frequency is due to the selection of the different RF

preamplifiers on different bands. The S-meter calibration of both radios was independent of mode. The image and various IF rejections of both radios was extremely good, generally in excess of 110dB. Some low-level inband 'birdies' were audible on the FT-1000MP, particularly on 24MHz.
The close-in reciprocal mixing performance of the IC-775DSP was extremely good, very similar to the IC-737 and about the best I have measured on any radio. The FT-1000MP was poorer in this respect but excelled in terms of close-in two-tone dynamic range. These two parameters are shown plotted in Fig 1. The strange shape of the FT-1000MP two-tone dynamic range curve is probably due to intermodulation occurring in more than one place, giving rise to cancellation and artificially high figures. The overall figures for intercept and dynamic range are good but are equalled and bettered in some cases by other radios on the market. The IF skirt selectivities of the Yaesu filters are extremely good. The Icom radio was not fitted with the optional filters. Measurements of second order intermodulation, which can be troublesome with strong broadcast stations eg 7.1 + 7.2 = 14.3MHz, showed an average result for the IC-775DSP but the FT-1000MP was some 10dB better than any other radio measured. This is probably due to the use of a larger number (11) of narrow bandwidth frontend filters in this radio, particularly in the critical mid-frequency region.

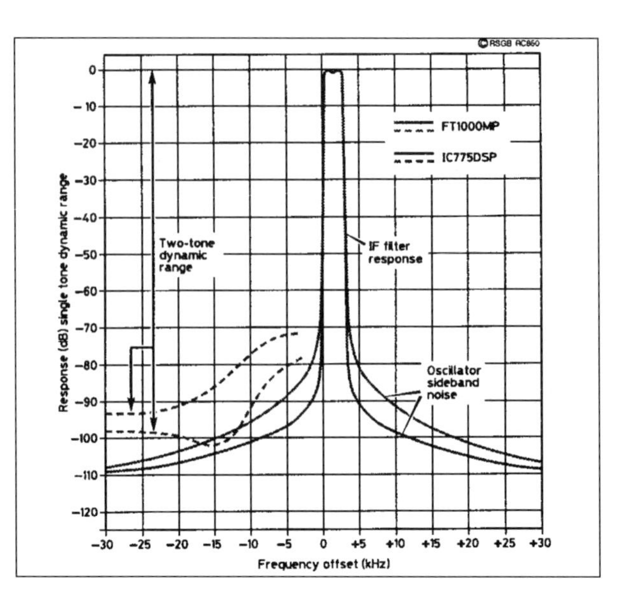


*Fig 1: Effective selectivity curves on USB of FT-1000MP and IC-775DSP.*

## *TRANSMITTER MEASUREMENTS*

Reducing power on the IC-775DSP to 100 - 150W made a big improvement in transmitter intermodulation products on SSB. In both cases the auto-ATU introduced a loss of about 10% in power output. The keying waveforms of both radios showed low distortion and reasonably shaped rise and fall times. The first character in semi break-in mode on the IC-775DSP was noticeably clipped but OK in full break-in mode.

## ON-THE-AIR PERFORMANCE

BOTH THE IC-775DSP and FT-1000MP are highly sophisticated radios, and during the period of the review it was difficult to grasp fully all aspects of all the features provided. Both radios gave impeccable performance on receive and on transmit. Signal handling was excellent and the various selectivity functions all performed very well. The DSP enhancements on both radios were most effective, particularly the noise reduction and auto-notch features. The transmit audio adjustments were effective and should accommodate virtually any microphone characteristic.
Ergonomically, both radios were easy to use, but I think the FT-1000MP just had the edge. IF filter selection on the IC-775DSP is less obvious than the FT-1000MP and the quick split did not always seem to perform as expected. The CW sidetone on the Icom was poor with clicks and shortened characters. The transmitted signal, however, was fine. The higher transmit power of the Icom can be a distinct advantage but if driving a linear, the extra power is not needed and care should be taken not to

overdrive. The lower distortion at the reduced power setting can be a definite advantage. The only negative comments on the FT-1000MP are the fan, which is slightly noisy, and the unlit display segments which are rather visible. VOX needs to be engaged to select CW semi break-in. This can then result in tripping on SSB but can be avoided by reducing the VOX gain to zero. Naturally this disables SSB VOX - which I never use in any case.

## CONCLUSIONS

BOTH OF THESE RADIOS are real top performers for DX and contest working. The performance of both are very similar, some parameters are slightly better on the IC-775DSP and others on the FT-1000MP. In terms of ergonomics my preference is marginally for the Yaesu and the Yaesu has possibly more features. The second receiver is certainly better implemented on the Yaesu. The Icom, however, has double the transmit power.

The recommended retail price of the IC-775DSP is £3699 and of the FT-1000MP £2849.

## ACKNOWLEDGEMENTS

I WOULD LIKE TO thank Icom (UK) of Herne Bay, Kent for the loan of the IC-775DSP and Yaesu UK for the loan of the FT-1000MP.

## YAESU FT-1000MP MEASURED PERFORMANCE

### RECEIVER MEASUREMENTS

| FREQUENCY | SENSITIVITY SSB 10dB s+n:n PREAMP IN | SENSITIVITY SSB 10dB s+n:n PREAMP OUT | INPUT FOR S9 PREAMP IN | INPUT FOR S9 PREAMP OUT |
|---|---|---|---|---|
| 1.8 MHz | 0.32μV (-117dBm) | 0.40μV (-115dBm) | 200μV | 250μV |
| 3.5 MHz | 0.40μV (-115dBm) | 0.35μV (-116dBm) | 250μV | 250μV |
| 7 MHz | 0.32μV (-117dBm) | 0.35μV (-116dBm) | 200μV | 250μV |
| 10 MHz | 0.18μV (-122dBm) | 0.40μV (-115dBm) | 100μV | 280μV |
| 14 MHz | 0.20μV (-121dBm) | 0.40μV (-115dBm) | 110μV | 320μV |
| 18 MHz | 0.16μV (-123dBm) | 0.35μV (-116dBm) | 100μV | 320μV |
| 21 MHz | 0.16μV (-123dBm) | 0.40μV (-115dBm) | 90μV | 320μV |
| 24 MHz | 0.13μV (-125dBm) | 0.45μV (-114dBm) | 35μV | 350μV |
| 28 MHz | 0.13μV (-125dBm) | 0.50μV (-113dBm) | 25μV | 400μV |

| S-READING (14MHz) | INPUT LEVEL SSB (PREAMP OFF) |
|---|---|
| S1 | 4.5μV |
| S3 | 6.3μV |
| S5 | 11μV |
| S7 | 40μV |
| S9 | 250μV |
| S9+20 | 2.5mV |
| S9+40 | 20mV |
| S9+60 | 180mV |

AM sensitivity (28MHz): 0.7μV for 10dB s+n:n at 30% mod depth

FM sensitivity (28MHz): 0.13μV for 12dB SINAD 3kHz pk deviation

AGC threshold: 2.5μV

100dB above AGC threshold for +2dB audio output

AGC attack time: 5ms (fast), 2ms (slow)

AGC decay time: 0.2-0.5s (fast), 2-3s (slow)

Max audio before clipping: 8Ω-1.6W, 4Ω-2.7W at 1% distortion

Inband intermodulation products: -26 to -40dB

| IF FILTER IF2/IF3 | IF BANDWIDTH -6dB | IF BANDWIDTH -60dB |
|---|---|---|
| SSB 2.4/2.4 | 2590Hz | 3420Hz |
| SSB 2.0/2.0 | 2000Hz | 2560Hz |
| CW 500/500 | 480Hz | 880kHz |
| CW 250/250 | 270Hz | 450Hz |
| AM THRU/6K | 6370Hz | 13.1kHz |
| FM | 6370Hz | 13.1kHz |

INTERMODULATION (50kHz Tone Spacing)

| Frequency | PREAMP IN 3rd order intercept | PREAMP IN 2 tone dynamic range | PREAMP OUT 3rd order intercept | PREAMP OUT 2 tone dynamic range |
|---|---|---|---|---|
| 1.8 MHz | +13dBm | 93dB | +15dBm | 93dB |
| 3.5 MHz | +18dBm | 95dB | +19dBm | 97dB |
| 7 MHz | +19dBm | 97dB | +20dBm | 97dB |
| 14 MHz | +19dBm | 99dB | +29dBm | 102dB |
| 21 MHz | +14dBm | 97dB | +19dBm | 96dB |
| 28 MHz | -4dBm | 87dB | +6dBm | 86dB |

CLOSE-IN INTERMODULATION ON 7MHz BAND

| Spacing | PREAMP IN 3rd order intercept | PREAMP IN 2 tone dynamic range | PREAMP OUT 3rd order intercept | PREAMP OUT 2 tone dynamic range |
|---|---|---|---|---|
| 3 kHz | -11dBm | 77dB | -8dBm | 78dB |
| 5 kHz | -8dBm | 79dB | -2dBm | 82dB |
| 7 kHz | -2dBm | 83dB | +1dBm | 84dB |
| 10 kHz | +10dBm | 91dB | +13dBm | 92dB |
| 15 kHz | +22dBm | 99dB | +29dBm | 103dB |
| 20 kHz | +22dBm | 99dB | +24dBm | 99dB |
| 30 kHz | +19dBm | 97dB | +22dBm | 98dB |
| 40 kHz | +19dBm | 97dB | +21dBm | 97dB |
| 50 kHz | +19dBm | 97dB | +20dBm | 97dB |

| FREQUENCY OFFSET | RECIPROCAL MIXING FOR 3dB NOISE | BLOCKING | TX NOISE IN 2.5kHz BANDWIDTH |
|---|---|---|---|
| 3 kHz | 80dB | -25dBm | -76dBC |
| 5 kHz | 86dB | -25dBm | -81dBC |
| 10 kHz | 94dB | -18dBm | -89dBC |
| 15 kHz | 99dB | -10dBm | -94dBC |
| 20 kHz | 103dB | -6dBm | -96dBC |
| 30 kHz | 108dB | -4dBm | -99dBC |
| 50 kHz | 115dB | 0dBm | -101dBC |
| 100 kHz | 121dB | 0dBm | -103dBC |
| 200 kHz | 124dB | 0dBm | -104dBC |

TRANSMITTER MEASUREMENTS

| FREQUENCY | CW POWER OUTPUT | SSB (PEP) POWER OUTPUT | HARMONICS | INTERMODULATION PRODUCTS 3rd order | INTERMODULATION PRODUCTS 5th order |
|---|---|---|---|---|---|
| 1.8 MHz | 98W | 104W | -64dB | -26dB | -40dB |
| 3.5 MHz | 97W | 103W | -65dB | -26dB | -40dB |
| 7 MHz | 96W | 101W | -62dB | -25dB | -44dB |
| 10 MHz | 95W | 101W | -66dB | -26dB | -44dB |
| 14 MHz | 96W | 102W | -65dB | -24dB | -40dB |
| 18 MHz | 97W | 101W | -52dB | -24dB | -40dB |
| 21 MHz | 97W | 101W | -70dB | -22dB | -40dB |
| 24 MHz | 98W | 102W | -65dB | -20dB | -37dB |
| 28 MHz | 98W | 102W | -70dB | -25dB | -35dB |

**Carrier suppression:** >80dB; **Sideband suppression:** >80dB @ 1kHz; **Transmitter noise:** see table above; **Transmitter AF distortion:** 0.5%; **Microphone input sensitivity:** 1mV for full output; **SSB T/R switch speed:** mute-TX 12ms, TX-mute 5ms, mute-RX 24ms, RX-mute 2ms.

**NOTE:** All signal input voltages given as PD across antenna terminal. Unless stated otherwise, all measurements made on SSB with the receiver preamp switched in. All two-tone transmitter intermodulation products quoted with respect to either originating tone.

REVIEWS

## ICOM IC-775DSP MEASURED PERFORMANCE

### RECEIVER MEASUREMENTS

| FREQUENCY | SENSITIVITY SSB 10dB s+n:n PREAMP IN | SENSITIVITY SSB 10dB s+n:n PREAMP OUT | INPUT FOR S9 PREAMP IN | INPUT FOR S9 PREAMP OUT |
|---|---|---|---|---|
| 1.8 MHz | 0.13μV (-125dBm) | 0.25μV (-119dBm) | 32μV | 89μV |
| 3.5 MHz | 0.13μV (-125dBm) | 0.22μV (-120dBm) | 32μV | 89μV |
| 7 MHz | 0.13μV (-125dBm) | 0.22μV (-120dBm) | 25μV | 71μV |
| 10 MHz | 0.11μV (-126dBm) | 0.20μV (-121dBm) | 25μV | 71μV |
| 14 MHz | 0.14μV (-124dBm) | 0.25μV (-119dBm) | 25μV | 71μV |
| 18 MHz | 0.14μV (-124dBm) | 0.25μV (-119dBm) | 25μV | 71μV |
| 21 MHz | 0.14μV (-124dBm) | 0.28μV (-118dBm) | 28μV | 89μV |
| 24 MHz | 0.14μV (-124dBm) | 0.28μV (-118dBm) | 22μV | 89μV |
| 28 MHz | 0.16μV (-123dBm) | 0.28μV (-118dBm) | 25μV | 89μV |

| S-READING (14MHz) | INPUT LEVEL SSB (PREAMP OFF) |
|---|---|
| S1 | 4.8μV |
| S3 | 6.3μV |
| S5 | 11μV |
| S7 | 24μV |
| S9 | 71μV |
| S9+20 | 710μV |
| S9+40 | 7.1mV |
| S9+60 | 45mV |

AM sensitivity (28MHz): 0.9μV for 10dB s+n:n at 30% mod depth.

FM sensitivity (28MHz): 0.2μV for 12dB SINAD 3kHz pk deviation.

AGC threshold: 3.5μV.

100dB above AGC threshold for +0.3dB audio output.

AGC attack time: 2ms.

AGC decay time: 0.2-7s.

Max audio before clipping: 2.1W into 8Ω.

Inband intermodulation products: -32 to -40dB.

| MODE | IF BANDWIDTH -6dB | IF BANDWIDTH -60dB |
|---|---|---|
| SSB | 2540Hz | 3350Hz |
| CW(500)470Hz | 1120Hz | |
| AM | Not measured | |
| FM | Not measured | |

INTERMODULATION (50kHz Tone Spacing)

| Frequency | PREAMP IN 3rd order intercept | PREAMP IN 2 tone dynamic range | PREAMP OUT 3rd order intercept | PREAMP OUT 2 tone dynamic range |
|---|---|---|---|---|
| 1.8 MHz | +3dBm | 91dB | +9dBm | 92dB |
| 3.5 MHz | +4dBm | 92dB | +11dBm | 93dB |
| 7 MHz | 0dBm | 90dB | +10dBm | 93dB |
| 14 MHz | +3dBm | 91dB | +12dBm | 93dB |
| 21 MHz | +3dBm | 91dB | +12dBm | 93dB |
| 28 MHz | -8dBm | 83dB | +1dBm | 86dB |

CLOSE-IN INTERMODULATION ON 7MHz BAND

| | PREAMP IN | | PREAMP OUT | |
|---|---|---|---|---|
| Spacing | 3rd order intercept | 2 tone dynamic range | 3rd order intercept | 2 tone dynamic range |
| 3 kHz | -31dBm | 69dB | -22dBm | 72dB |
| 5 kHz | -31dBm | 69dB | -22dBm | 72dB |
| 7 kHz | -30dBm | 70dB | -21dBm | 73dB |
| 10 kHz | -23dBm | 75dB | -13dBm | 78dB |
| 15 kHz | -9dBm | 84dB | 0dBm | 87dB |
| 20 kHz | -4dBm | 87dB | +7dBm | 91dB |
| 30 kHz | 0dBm | 90dB | +10dBm | 93dB |
| 40 kHz | 0dBm | 90dB | +10dBm | 93dB |
| 50 kHz | 0dBm | 90dB | +10dBm | 93dB |

| FREQUENCY OFFSET | RECIPROCAL MIXING FOR 3dB NOISE | BLOCKING | TX NOISE IN 2.5kHz BANDWIDTH |
|---|---|---|---|
| 3 kHz | 90dB | -24dBm | -78dBC |
| 5 kHz | 96dB | -24dBm | -80dBC |
| 10 kHz | 100dB | -19dBm | -87dBC |
| 15 kHz | 103dB | -13dBm | -92dBC |
| 20 kHz | 106dB | -5dBm | -97dBC |
| 30 kHz | 109dB | -5dBm | -100dBC |
| 50 kHz | 113dB | -5dBm | -101dBC |
| 100 kHz | 117dB | -5dBm | -102dBC |
| 200 kHz | 120dB | -5dBm | -102dBC |

TRANSMITTER MEASUREMENTS

| FREQUENCY | CW POWER OUTPUT | SSB (PEP) POWER OUTPUT | HARMONICS | INTERMODULATION PRODUCTS 3rd order | INTERMODULATION PRODUCTS 5th order |
|---|---|---|---|---|---|
| 1.8 MHz | 218W | 215W | -65dB | -24dB | -38dB |
| 3.5 MHz | 210W | 210W | -68dB | -32dB | -35dB |
| 7 MHz | 208W | 208W | -65dB | -30dB | -35dB |
| 10 MHz | 208W | 207W | -70dB | -30dB | -36dB |
| 14 MHz | 205W | 205W | -70dB | -26dB | -34dB |
| 18 MHz | 204W | 205W | -70dB | -26dB | -32dB |
| 21 MHz | 203W | 204W | -70dB | -26dB | -32dB |
| 24 MHz | 204W | 203W | -70dB | -22dB | -34dB |
| 28 MHz | 203W | 202W | -65dB | -20dB | -32dB |

**Carrier suppression:** 60dB; **Sideband suppression:** 70dB @ 1kHz; **Transmitter noise:** see table above; **Transmitter AF distortion:** <0.5%; **Microphone input sensitivity:** 2mV for full output; **SSB T/R switch speed:** mute-TX 10ms, TX-mute 3ms, mute-RX 18ms, RX-mute 1ms.

**NOTE:** All signal input voltages given as PD across antenna terminal. Unless stated otherwise, all measurements made on SSB with the receiver preamp switched in. All two-tone transmitter intermodulation products quoted with respect to either originating tone.

# Kenwood TS-870 HF DSP Transceiver

### *by Peter Hart, G3SJX*

THE HOT topic of conversa tion in the world of HF at present is the appearance of new DSP rigs from Yaesu, Icom and Kenwood. The IC-775DSP and FT-1000MP were reviewed in the January 1996 issue of this magazine and this review of the TS-870S now completes the picture. Kenwood was the first to introduce digital signal processing (DSP) techniques into amateur HF transceivers with the TS-950SD during late 1989. Their latest offering, the TS-870S, is still one step ahead of the competition, as the first transceiver to implement full IF channel filtering using DSP. The consequences of this should become apparent during the course of this review. Ever since the January review appeared, I have been inundated with enquiries about the TS-870S and how it compares with the other DSP rigs. All will now be revealed, so read on.

## STANDARD FEATURES

THE TS-870S IS A 12V operated radio pitched as a successor to the TS-850S. Kenwood call this new radio 'Intelligent Digital Enhanced Communication System'. The receiver covers the frequency range from 30kHz to 30MHz and the transmitter is limited to the exact amateur allocations with regional variants available to cover the differences on 160, 80 and 40m in various parts of the world. Modes covered are USB, LSB, AM, FM, CW, reverse CW, FSK and reverse FSK. The reverse modes in effect use the opposite sideband which can be effective in combating adjacent channel interference. FSK is configurable for 170, 200, 425 or 850Hz shifts, high or low tone pairs and selectable polarity.

Tuning is in 10Hz steps at 10kHz per tuning knob revolution on SSB, CW and FSK, and 100Hz steps at 100kHz per knob revolution on AM and FM. These tuning rates may be optionally halved and a FINE button decreases step sizes by a factor of ten, giving 1Hz steps on CW / SSB / FSK. UP / DOWN keys scroll through the amateur bands or alternatively in 1MHz steps (or 100 / 500kHz). Only a single band store is provided and this returns the last used frequencies, mode, antenna and various other settings on any particular band. A click-step rotary allows for rapid coarse frequency steering at 1 / 5 / 10kHz per step or 9kHz for AM broadcast, and this control also selects memories and menus as appropriate. 10 multifunction keys also allow for direct frequency input.

Twin VFOs are provided (A / B). In split mode there is a one touch transmit frequency set / monitor in split mode. 100 memories are included with all the usual facilities and a quick memory stack which provides a single keystroke store and recall of the last five settings on a last-in first-out basis. Comprehensive scanning across programmed limits or memory groupings is also provided. RIT and XIT functions over a range of ±9.99kHz in 10Hz or 1Hz steps, a larger range than the TS-850S and convenient for much split frequency DX working.

The receiver front end is switchable between NORMAL operation at full sensitivity and AIP (Advanced Intercept Point). AIP switches out the preamplifier and gives improved strong signal handling at the expense of reduced sensitivity. In NORMAL operation, a higher gain preamplifier is enabled above 21.49MHz. An additional 6, 12 or 18dB input attenuator may also be selected. Other receiver functions include fully variable AGC, analogue noise blanker in addition to DSP, all-mode squelch and CW pitch variable between 400Hz and 1000Hz.

The transmitter power is variable up to 100W and transmit features include speech processor, VOX and audio transmission monitor. Full and semi break-in is provided on CW, as is a fully fledged memory keyer. CTCSS tone access is included for FM repeaters. An automatic ATU is built-in which may be switched in circuit on receive as well as transmit.

The multicoloured display indicates main frequency to 10Hz resolution, memory channel, mode and a host of status indicators. A smaller numeric display indicates second frequency in split operation, RIT / XIT offset or IF filter bandwidth parameters according to context. Two 30-segment curved bargraph meters, emulating analogue meters, provide an S meter and a pictorial impression of receiver bandwidth. The S meter reading may be compensated to give the same reading with AIP on as well as off and on FM two styles of linearity are selectable. On transmit, one meter indicates transmit power and the other ALC, SWR or compression level. A peak hold function may also be enabled.

The connectors on the rear panel provide the usual interfaces to a linear amplifier, data terminals, CW keyer and keying paddles, IF output for SM-230 monitor scope, external receiver and PC control. There is no provision, however, for a separate antenna for the receiver.

Special Features

A FULLY FEATURED K-1 Logikey keyer is built into the TS-870S. As well as providing a very comprehensive keyer emulating various keyer types with variable weighting and a speed range of 6 - 60WPM, a host of useful contest functions are also included. Four message stores are provided up to 220 characters maximum total, selected by pushbuttons on the front panel. These may be queued and may call each other. It is possible to pause in a message for insertion of manual text and then automatically resume. Automatically incrementing contest serial numbers can be sent with or without leading zeros and with various options for abbreviated numbers (T and N for 0 and 9). The start number can be specified and decremented in case of errors. Programming and other commands are sent to the keyer in Morse code using the paddle, and messages may be programmed in real time or entered one word at a time with edit capability.

A similar unit for voice modes is available as an optional extra, the DRU-3 digital recording system. This will record up to four channels of maximum 15 seconds of audio per channel from the microphone for later transmission. The channels may be cascaded in any order and set to repeat after a programmable delay.

Twin antenna connectors are provided, selectable from the front panel and the selection is stored with frequency. Eighteen frequency ranges are identified, with the splits so arranged to allow separate antennas for the CW and SSB segments of most bands to be automatically selected if required.

A full RS-232C computer interface is provided which needs no external interface unit, and operates with selectable data rates from 1200 to 57,600 baud. Virtually every parameter in the radio is programmable under computer control, including all the menu, memory and DSP functions. It is also possible to link two transceivers together, and to other models in the Kenwood range and directly transfer operating frequency and mode.

Most radios these days have user-programmable settings for many of the features. This is very well implemented in the TS-870S, where some 68 functions are programmable, simultaneously with the radio in full operation. Hence the immediate effect of setting say the audio bandwidth on transmit can be heard on the transmitted signal. These menu functions include many of the DSP characteristics, step sizes, reprogramming of front panel controls, FSK parameters etc. Two complete sets of menu settings may be stored to allow two different operator preferences or home and field day use for instance.

DSP

THE FEATURE WHICH sets this radio apart from all others is the DSP unit. On receive, this is used to provide high grade IF filtering for the main selectivity, demodulation in all modes, audio filtering, AGC, notch filters and noise reduction. On transmit the DSP unit provides modulation, audio filtering, speech compression, equalisation, microphone AGC, VOX and on CW, carefully shaped envelope rise and fall times. The IF filters are implemented at the final IF of 11.3kHz. On SSB, independent control of the low frequency cutoff from 0 to 1000Hz in 10 steps, and the high frequency cutoff from 1.4kHz to 6kHz in 12 steps provides all the facilities of variable bandwidth, IF shift, passband tuning or slope tuning. Four low frequency and six high frequency cutoffs are provided on AM, six bandwidth settings (5 - 14kHz) on FM and four bandwidth settings on FSK. On CW, there are six bandwidth settings of 50, 100, 200, 400, 600 and 1000Hz. Filter ringing is minimal, even with 50Hz IF bandwidth.

Two auto-notch filters are provided. An IF auto-notch will eliminate and track a single heterodyne without it affecting the AGC and hence reducing sensitivity. Beat cancel is an audio function which will remove multiple heterodynes but after the AGC has acted.

Two noise reduction methods may be selected. The Line Enhancer Method (LEM) is particularly suited to SSB and is based on adaptive filtering. The Speech Processing / Auto Correlation (SPAC) method uses a statistical auto-correlation algorithm and is claimed to be ideal for CW.

The transmit audio bandwidth may be tailored for low and high frequency cutoff, high and low frequency boost or cut for the speech processor and a separate equaliser giving treble boost, bass boost or comb filter response. The comb filter is intended for removing background ambient noise. There should be a combination to suit virtually every microphone and circumstance. The speech processor processes the audio in three sub-bands. This reduces the amount of distortion generated.

The DSP section is very powerful, using two Motorola DSP56002FC40 24-bit processors, each running at 20MIPS (20 million instructions per second). Interfacing to the analogue world is via 18-bit A / D and 16-bit D / A converters giving overall dynamic ranges of around 100dB. A custom gate array, programmable logic, fast SRAM and 768kb of program memory completes the DSP line-up.

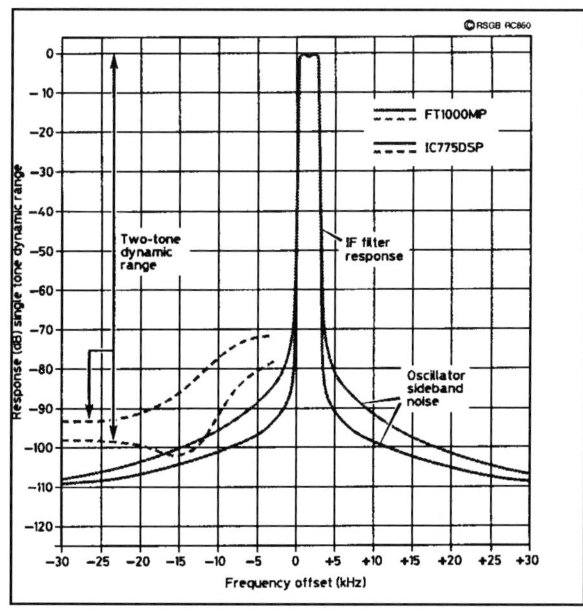


*Fig 1: Effective selectivity curve of the TS-870S on USB.*

## DESCRIPTION

THE TS-870S IS similar in size and weight to the earlier TS-850S, measuring 33.9W x 13.5H x 37.5Dcm, and weighing around 11kg. The construction is conventional.

The receiver uses a quadruple conversion superhet architecture with IFs of 73.05MHz, 8.83MHz, 455kHz and 11.3kHz. My first thought was why so many conversions? This is normally done to implement passband tuning, IF shift and IF notch and these facilities are now all implemented by the DSP at 11.3kHz. However, accepting that the first IF needs to be high to enable general coverage and ease filtering, if only one further conversion is used to 11.3kHz this would require high grade and hence high cost filtering to eliminate images. Two intermediate conversions allow cheap filters to be used. Although the main selectivity is provided by DSP, switchable 3kHz, 6kHz and 15kHz bandwidth filters are also fitted at both 8.83MHz and 455kHz IFs as roofing filters to ease out-of-band problems at the DSP.

The receiver front end uses one of twelve input bandpass filters switched by PIN diodes. Narrow bandwidth filters are used on 7, 14 and 21MHz to help protect against intermodulation. The transmitted signal is generated by the DSP at 11.3kHz and mixed through the 455kHz, 8.83MHz and 73.05MHz IFs to final frequency.

## MEASUREMENTS

THE FULL MEASUREMENTS are detailed in the tables, with additional comments as follows.

### RECEIVER MEASUREMENTS

The S meter linearity is good and the same calibration the applies to all modes. The

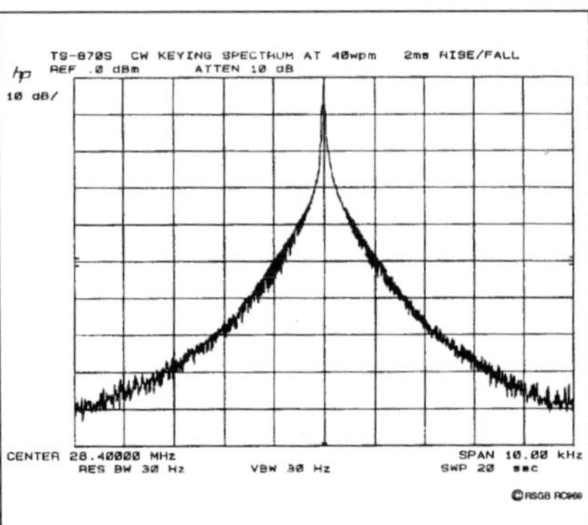


Fig 2: TS870S CW keying spectrum at 40WPM:2ms rise/fall time.

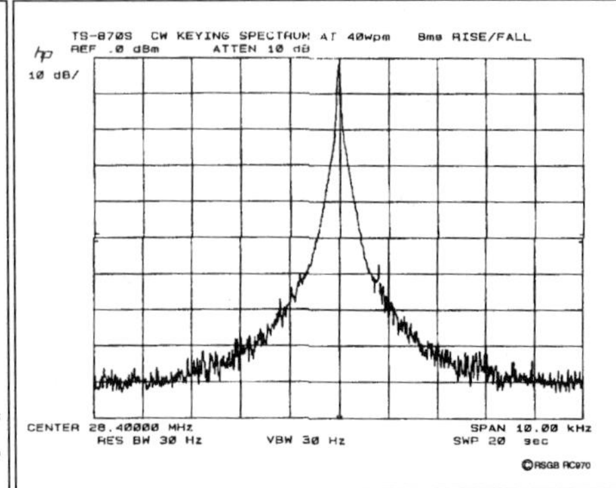


Fig 3: TS870S CW keying spectrum at 40WPM:8ms rise/fall time.

rejection of all image, IF and other spurious responses was excellent, at better than 100dB. The only exception was rejection of the 8.83MHz IF on 10MHz, which was only 75dB. The IF selectivity measurements were a revelation. The 6 / 60dB shape factors are superior to any crystal filters measured to date, and the passband response was absolutely flat. Crystal filters always have several ripples of up to + / - 2db across the passband. The strong signal performance in terms of intermodulation, blocking and reciprocal mixing at greater than 15kHz off-tune is very good. However, the intermodulation and blocking characteristic degrades very badly closer in, giving only 58dB dynamic range at 3kHz offset. This is due to signal handling problems at the 2nd or 3rd IF and is a big danger, if not an inevitable consequence, of putting the main selectivity at the back end of the signal RF path. **Fig 1** shows the overall effect on selectivity of the various parameters.

## TRANSMITTER MEASUREMENTS

The transmitter power output figures are given for the ATU out of circuit. In circuit figures are about 10% less. The power output was reducible to about 10W, and the power meter bargraph was accurate to about 10%.

The transmitter audio distortion was very low although some non-harmonic in-band products were seen with two-tone audio drive. The speech processor was very clean.

CW full and semi break-in gave similar results. The distortion was low at 40WPM and there was negligible clipping of first characters. The different rise / fall times gave a marked difference to the signal spectrum as shown in **Fig 2 and Fig 3**. 8ms rise / fall is entirely adequate at speeds of over 40WPM and gives up to 20dB reduction in adjacent channel sidebands compared with rather poor 2ms result. Note that I have measured somewhat better results for the TS-950SD with DSP shaping.

## *ON THE AIR*

SEVERAL ASPECTS OF the receiver performance are immediately apparent. The audio quality is excellent on all modes, particularly if the bandwidth is opened up. The internal speaker does not really do this full justice and results are better on headphones or external speaker. The IF filter response is really excellent. Signals come and go very cleanly on tuning, and control and flexibility are very good. My only complaint is the display which fails to continuously indicate filter status. It is necessary to change the relevant parameter (eg width) to enable the display. Extra intermediate CW bandwidths would also be an advantage, eg 300Hz and 500Hz. The narrow CW filters were most impressive with a lack of the normal ringing experienced.

The main problem experienced with this receiver was close-in signal handling. This was particularly noticed on

160m DX working. With two large CW signals on, say, 1830 and 1833kHz, bits of disjointed CW would be heard on 1827 and 1836kHz. This lost me several DX contacts. Similarly during the 80m CW AFS contest, rarely did I have a quiet background noise level although I had, in theory, superb IF filtering flexibility at my fingertips.

The noise reduction capabilities were interesting and effective in different situations. However, don't expect it to produce signals which could otherwise not be heard. SPAC could produce some very strange-sounding audio on SSB, although it was very effective on CW. There was a slight hint of roughness when tuning at speed but this disappeared at slower tuning rates.

The transmit audio quality was excellent, real broadcast quality if the bandwidth was opened up or could be tailored for good punchy DX working needs. There is tremendous flexibility provided. On PTT, there was very occasionally a slight delay returning to receive for some reason. The reason was unknown. CW clicks were apparent with 2ms and 4ms rise / fall times, but not at the higher settings. Use 8ms at all times please. The fan was very quiet and never noticed.

## CONCLUSIONS

THE INCLUSION OF DSP in the TS-870S gives some very impressive results with selectivity and audio quality unsurpassed in any other radio. Unfortunately, the close-in dynamic range is particularly poor, but perhaps this is insurmountable with the DSP architecture adopted. In most situations this will not be of great consequence, but for top-flight DXing and contest working, primarily on the low bands, this needs to be taken into consideration.

The current list price is £2399.95 inc VAT. Remember that no optional filters are needed, this can easily add £500 to the cost of a conventional radio. I would like to thank Trio-Kenwood UK Ltd for the loan of the equipment.

## KENWOOD TS-870S MEASURED PERFORMANCE

### RECEIVER MEASUREMENTS

| FREQUENCY | SENSITIVITY SSB 10dBs+n:n PREAMP IN | SENSITIVITY SSB 10dBs+n:n PREAMP OUT | INPUT FOR S9 PREAMP IN | INPUT FOR S9 PREAMP OUT |
|---|---|---|---|---|
| 1.8 MHz | 0.16μV (-123dBm) | 0.45μV (-114dBm) | 28μV | 110μV |
| 3.5 MHz | 0.13μV (-125dBm) | 0.35μV (-116dBm) | 28μV | 110μV |
| 7 MHz | 0.18μV (-122dBm) | 0.63μV (-111dBm) | 35μV | 160μV |
| 10 MHz | 0.14μV (-124dBm) | 0.40μV (-115dBm) | 28μV | 110μV |
| 14 MHz | 0.16μV (-123dBm) | 0.50μV (-113dBm) | 32μV | 140μV |
| 18 MHz | 0.14μV (-124dBm) | 0.40μV (-115dBm) | 28μV | 100μV |
| 21 MHz | 0.16μV (-123dBm) | 0.63μV (-111dBm) | 35μV | 160μV |
| 24 MHz | 0.10μV (-127dBm) | 0.50μV (-113dBm) | 18μV | 140μV |
| 28 MHz | 0.10μV (-127dBm) | 0.50μV (-113dBm) | 28μV | 140μV |

| S-READING (14MHz) | INPUT LEVEL SSB PREAMP IN | INPUT LEVEL SSB PREAMP OUT |
|---|---|---|
| S1 | 1.3μV | 5.6μV |
| S3 | 1.8μV | 9μV |
| S5 | 3.2μV | 14μV |
| S7 | 9μV | 40μV |
| S9 | 32μV | 140μV |
| S9+20 | 320μV | 1.4mV |
| S9+40 | 2.8mV | 13mV |
| S9+60 | 22mV | 90mV |

| IF FILTER | IF BANDWIDTH -6dB | IF BANDWIDTH -60dB |
|---|---|---|
| SSB 2.3kHz | 2310Hz | 3150Hz |
| AM 6kHz | 12.0kHz | 15.0kHz |
| FM 14kHz | 13.6kHz | 16.6kHz |
| CW 600Hz | 630Hz | 1025Hz |
| CW 400Hz | 415Hz | 705Hz |
| CW 200Hz | 210Hz | 380Hz |
| CW 100Hz | 115Hz | 302Hz |
| CW 50Hz | 68Hz | 226Hz |

**AM sensitivity (28MHz):** 0.7μV for 10dBs+n:n at 30% mod depth

**FM sensitivity (28MHz):** 0.22μV for 12dB SINAD 3kHz pk deviation

**AGC threshold:** 1μV

100dB above AGC threshold for +0.5dB audio output

**AGC attack time:** 2ms

**AGC decay time:** 0.1 - 7s

**Max audio before clipping:** 1.8W into 8Ω at <1% distortion

**Inband intermodulation products:** -36 to -40dB

INTERMODULATION (50kHz Tone Spacing)

| Frequency | PREAMP IN 3rd order intercept | PREAMP IN 2 tone dynamic range | PREAMP OUT 3rd order intercept | PREAMP OUT 2 tone dynamic range |
|---|---|---|---|---|
| 1.8 MHz | +2dBm | 90dB | +11dBm | 90dB |
| 3.5 MHz | +2dBm | 91dB | +12dBm | 92dB |
| 7 MHz | +5dBm | 91dB | +22dBm | 95dB |
| 14 MHz | +3dBm | 91dB | +17dBm | 93dB |
| 21 MHz | +5dBm | 92dB | +22dBm | 95dB |
| 28 MHz | -6dBm | 87dB | +20dBm | 95dB |

CLOSE-IN INTERMODULATION ON 7MHz BAND

| Spacing | PREAMP IN 3rd order intercept | PREAMP IN 2 tone dynamic range | PREAMP OUT 3rd order intercept | PREAMP OUT 2 tone dynamic range |
|---|---|---|---|---|
| 3 kHz | -45dBm | 58dB | -34dBm | 58dB |
| 5 kHz | -42dBm | 60dB | -29dBm | 61dB |
| 7 kHz | -39dBm | 62dB | -22dBm | 66dB |
| 10 kHz | -28dBm | 69dB | -12dBm | 73dB |
| 15 kHz | -9dBm | 82dB | +5dBm | 84dB |
| 20 kHz | +4dBm | 91dB | +20dBm | 94dB |
| 30 kHz | +5dBm | 91dB | +22dBm | 95dB |
| 40 kHz | +5dBm | 91dB | +22dBm | 95dB |
| 50 kHz | +5dBm | 91dB | +22dBm | 95dB |

## TRANSMITTER MEASUREMENTS

| FREQUENCY | CW POWER OUTPUT | SSB(PEP) POWER OUTPUT | HARMONICS | INTERMODULATION PRODUCTS 3rd order | INTERMODULATION PRODUCTS 5th order |
|---|---|---|---|---|---|
| 1.8 MHz | 110W | 125W | -68dB | -28dB | -40dB |
| 3.5 MHz | 107W | 120W | -70dB | -30dB | -40dB |
| 7 MHz | 106W | 115W | -64dB | -25dB | -45dB |
| 10 MHz | 105W | 115W | -65dB | -26dB | -41dB |
| 14 MHz | 105W | 115W | -72dB | -25dB | -41dB |
| 18 MHz | 104W | 112W | -70dB | -20dB | -38dB |
| 21 MHz | 102W | 110W | -63dB | -30dB | -32dB |
| 24 MHz | 102W | 107W | -68dB | -15dB | -32dB |
| 28 MHz | 100W | 103W | -75dB | -15dB | -30dB |

**Carrier suppression:** 60dB; **Sideband suppression:** >80dB @ 1kHz; **Transmitter noise:** see table above; **Transmitter AF distortion:** <1%; **Microphone input sensitivity:** 1mV for full output; **SSB T/R switch speed:** mute-Tx 13ms, Tx-mute 3ms, mute-Rx 15ms, Rx-mute 2ms.

**NOTE:** Preamp out corresponds to AIP on. Measurements made with the DSP noise reduction facilities switched out. All signal input voltages given as PD across antenna terminal. Unless stated otherwise, all measurements made on SSB with the receiver preamp switched in with IF bandwidth set to 300 - 2600Hz. All two-tone transmitter intermodulation products quoted with respect to either originating tone.

# IC-706 Eleven Band Transceiver

## *A user review by RSGB HQ Staff*

TAKE A LOOK around your shack. You've probably got most of the following: an HF transceiver, a 6m transceiver, a 2m multi-mode, a general coverage receiver, a VHF scanner, an electronic keyer, a speech compressor and an AM/FM broadcast radio. What if you could replace the whole lot with one box retailing at under £1200? And what if the box were so small you could fit it all in the car or in your holiday hand-luggage? A pipe dream? Not at all with the latest from Icom.

### *FACILITIES*

THE IC-706 MEASURES just 167W x 58H x 200Dmm and weighs in at 2.5kg. In this package, you get 100W on all HF bands, 100W on 50MHz and 10W on 144MHz, together with a receiver covering long wave to 200MHz, and whole lot more. The front panel detaches for remote operation using the optional separation cable.

Included with the radio is a sturdy double-fused 12V power lead, a hand micro-phone, spare fuses, and plugs for connecting a TNC, RTTY equipment, Morse key etc.

There are over 100 memories. 99 can contain independent transmit and receive frequen-cies, mode, CTCSS frequency or 1750Hz burst, and an eleven character name (eg 'Repeater R0' or 'Club net'). Two more contain scanning limits and another - the Call Channel - which is available on 144MHz only. The dial is not automatically locked during memory mode so once a memory is selected it performs in a similar (but not identical) way to VFO mode - this can lead to confusion.

Two VFOs are provided and memory contents can easily be transferred to either VFO - handy for rapidly switching between, say, 1.831MHz CW, 14.333MHz USB and 144.750 NBFM with repeater offset and CTCSS. Tuning steps are variable between 1Hz (200Hz per knob revolution) and 1MHz.

Up to ten Memo Pads are provided for the temporary storage of frequencies and modes (for instance of contest multiplier stations you are waiting for). These can be recalled at any time.

The acid test for many DX operators is "how easy is it to work split?". The answer is: "very easy". The 'quick split' option - pressing a function key for about a second - equalises the VFOs (or selects a programmable offset) and activates split operation. The transmit frequency can be monitored by a single button push. The word SPLIT appears prominently on the display to remind the operator to re-set after use.

For repeater operation it is possible to use the main tuning dial, having pro-grammed in a -600kHz offset (using the quick split function described above) and a 1750Hz toneburst. However it is much more convenient to use the memories because individual CTCSS tones can be set and the quick split function can be retained for HF use.

### *CONTROLS*

THE FRONT PANEL is crowded but, with the exception of the RIT button, is very accessible. The tuning knob, which occupies almost the whole height of the panel, includes a finger-hole and is a pleasure to use, even at the slowest setting.

Rotary controls are fitted for AF Gain and either Squelch or RF Gain, RIT shift and IF Shift. A miniature jack socket is provided for headphones.

All of the other controls are push buttons, the majority of which operate in conjunction with menus on the display. These handle facilities such as pre-amp/attenuator, memories, noise blanker etc.

Mode changing is simply a matter of pressing the Mode button the appropriate number of times. Band changing is achieved by pressing the TS button to select the band option, and then using the tuning knob.

The display is backlit bright yellow which makes the black LCD letters very easy to read. Many of the displayed items are user-selectable or only appear when the appropriate menu is accessed so the resultant display is not over-crowded. Visible all the time are the mode in use, 'Split' when accessed, a clear readout of the entire frequency (to 1Hz if selected), options such as 'NB' for noise blanker, the VFO in use, the memory in use, a bargraph and the current menu.

The rear panel is very

crowded. It contains a large heat-sink, two 'UHF' (SO-239) antenna sockets - one for below 60MHz and the other for above, a ground terminal, an accessory socket, jacks for external speaker, RTTY, remote control and key/paddle, a microphone socket (this is in parallel with one just below the front panel), 12V DC input and a socket for the optional automatic antenna tuner.

Usefully accessible through holes in the side panel are adjustments for speech compressor level, beep/ sidetone level, VOX and Anti-VOX.

### RECEIVER

THE RECEIVER is specified from 300kHz but the review model tuned down to 30kHz. It provides continuous coverage to 200MHz (automatically switching aerials at 60MHz) with relatively few 'birdies', none of which interfered with amateur bands reception. Note that this means yet another amateur band - 4m - is available, but only on receive.

Modes covered are USB/ LSB, CW, CW - reverse, RTTY (FSK), AM, NBFM and wide (broadcast) FM. Filter bandwidths are shown in the Specifications box opposite. Fitted as standard are additional narrow filters for AM and NBFM, the latter being useful for future 12.5kHz channel spacing or for operating on 29MHz. Disap-pointingly, a narrow CW filter is available only as an optional extra at £65 (500Hz) or £69 (250Hz).

Three pre-set levels of sensitivity are available from the front panel: normal, pre-amp and 20dB attenuator. These proved adequate for most purposes. However, if finer control is needed the Squelch knob can be set to act as an RF Gain control when used on CW, RTTY and SSB, whilst retaining its Squelch function on AM and NBFM.

A tuneable IF offset is provided to help reduce adjacent QRM. It comes with a novel miniature graphical display showing how the wanted signal relates to the filter edges.

The noise blanker proved effective on static and man-made noise in CW and SSB modes, but it completely ruined AM reception by badly distorting the wanted signal.

Tiny fingers are required to activate the RIT ON button. If care is taken not to 'nudge' the RIT Shift control, the RIT could be left on permanently. The button lights up to warn that RIT is on. A shift of up to 1kHz in either direction is permitted.

Slow or fast AGC time constants can be selected, but there is no facility to link AGC speed to the mode switch, so if fast AGC is selected for CW, slow AGC must be separately selected for SSB use.

The S-meter is an LED bargraph and, although no substitute for a real meter, it is well designed with fast attack, slow decay and an optional 'peak hold' facility.

Scanning is available either between programmed limits, on all memory channels or on selected memories. There are several methods of resuming the scan after a signal is detected.

In the short time the IC-706 was available for test, it was not possible to check one of the more esoteric facilities, the Simple Band Scope. This uses the dot matrix display to show graphically the activity over a band of frequencies.

The IC-706 was air-tested during the SAC CW Contest which provided a high level of activity. No problems were noticed once the RF gain was set to a sensible level. The receiver performed well on the medium wave with no cross-modulation, despite the high powered Brookmans Park BBC transmitter being only six miles away. One flaw emerged when tuning on the quiet 10m and 4m bands: The synthesiser was quite noisy, giving the impression of a busy band, until the knob was halted whereupon the 'signals' disappeared. The noise seemed to have been picked up on the long-wire antenna in use and would probably not have been prevalent on a remote coax-fed aerial.

An interesting facility is wide FM which allows the IC-706 to be used as a sensitive VHF broadcast receiver, albeit without stereo or squelch. This, together with the long and medium wave coverage, would allow the IC-706 to be used in a car in place of the broadcast set.

The loudspeaker was disappointing. It faces upwards, producing a good, clear, sound, but had a tendency to buzz on CW signals. Its frequency response was ideal for communications purposes but very poor for broadcast signals, even on AM. An external speaker sounded fine, though, and the volume could be turned up quite loud without noticeable distortion.

### TRANSMITTER

MODES AVAILABLE on transmit are: CW, USB/LSB, RTTY, AM and NBFM. Output is reduced to 40% on AM. The provision of 100W on 50MHz puts the IC-706 ahead of several of its rivals.

The supplied microphone felt good and has only the most basic controls - UP/DOWN buttons, a LOCK slide switch, and a lightly sprung transmit switch. The buttons can control frequency or memory channels.

Power output is continuously variable from a nominal 5W to 100W (1 to 10W on 2m). Transmit is inhibited outside amateur bands and Martin Lynch tells us that initial attempts to provide transmit on 4m have proved unsuccessful (looks like time for a bit of home-brewing!) The fan, which runs continuously - even on receive - speeds up noticeably on transmit.

The front panel bargraph can be switched to display relative power out, SWR or ALC. This latter is important when setting up the built-in speech compressor. VOX is available and adjustments can be made without opening up the box.

Icom have for some time incorporated an electronic keyer into their budget rigs, and the IC-706 is no exception. This means that yet another separate box has been done away with and greatly enhances the radio's value to the CW operator who enjoys portable working. Anyone who has forgotten their key on an expedition and has made QSOs by tapping wires together will appreciate the novel facility of being able to use the microphone's UP/DOWN keys as a Morse paddle - amazingly easy to use! In addition to taking a standard paddle (normal or reverse), the keyer can be configured to take a straight key or mechanical bug for those who prefer more personalised Morse.

Full or semi break-in is provided, together with 'no break-in' ie an external switch. Delay times are fully adjustable. Full break-in worked OK but the clattering of a relay on each dot proved very distracting. CW sidetone is adjustable from 300 to 900Hz.

### HANDBOOK

THE 60-PAGE manual is helpful and is written in reasonably good English. It describes all of the controls with helpful flow diagrams for the numerous Menu driven functions. It does, however, assume a knowledge of amateur radio terms and practice, unlike some other handbooks. Diagrams are used extensively and no problems were experienced in understanding the radio 's facilities. Usefully, Split and Repeater operation have a page each. No circuit or block diagrams are included.

### EXTRAS

OPTIONAL EXTRAS include a voice synthesiser (switchable English or Japanese), a high stability crystal unit, filters (250Hz, 500Hz, 2.8kHz or 1.9kHz), an automatic ATU, mobile mounting bracket, loudspeakers and many more. Note that, although the additional filters can be installed by the user (ie they are plug-in), only one at a time may be installed.

### CONCLUSION

THE IC-706 IS straightforward to drive and has a very usable front panel considering its size. On the air reports were all positive. And what fun to be able to listen to broadcast FM whilst automatically checking the local 2m net frequency every second or so!

Two things disappointed - the lack of a narrow CW filter and the synthesiser noise on 28 and 70MHz.

Having said that, this is a rig which can replace virtually all of your shack in one go at a price which is almost covered by the sale of your existing gear!

There's plenty to interest the keen Class B licensee, too, with 100W on 6m and a VHF general coverage receiver to check the progress of sporadic E. What is more, this really is a complete multiband station for taking mobile or on holiday. The IC-706 is set to become the ubiquitous rig of 1996. Now where's my cheque book . . .

### AVAILABLITY

THE IC-706 IS available from a number of *RadCom* advertisers (though currently demand appears to be exceeding supply). The review model was very kindly loaned by Martin Lynch - Amateur Radio Exchange Centre and we are most grateful to him and to the customer of his who was prepared to forego the pleasure of playing with his new toy for 48 hours whilst we checked it out.

REVIEWS

## SPECIFICATIONS

Source: IC-706 Handbook

| | |
|---|---|
| **GENERAL** | |
| Receive frequency coverage | 300kHz - 200MHz *(but see text)*. Specifications guaranteed only on amateur bands listed below. |
| Transmit frequency coverage (UK version) | 1.800 - 1.99999MHz; 3.500 - 3.9999MHz; 7.000 - 7.300MHz; 10.100 - 10.150MHz; 14.000 - 14.350MHz; 18.068 - 18.168MHz; 21.000 - 21.450MHz; 28.000 - 29.700MHz; 50.000 - 54.000MHz; 144.000 - 146.000MHz. |
| Modes | SSB, CW, AM, FM, WFM (receive only), RTTY. |
| Memory channels | 102 (split memory 99; scan edge 2; call channel 1). |
| Antenna impedance | 50Ω nominal. |
| Usable temperature range | -10°C to +60°C (+14°F to +140°F). |
| Frequency stability | Less than ±7ppm from 1 min to 60 min after power on.After that, rate of stability change is less than ±1ppm/hr at +25°C (+77°F). Temperature fluctuations 0°C to +50°C (+32°F to +122°F) less than ±5ppm. |
| Power supply requirement | 13.8V DC ±15%. |
| Current drain at 13.8V | Transmit 20A; Receive squelched 1.5A; Receive max audio 1.7A. |
| Dimensions | Millimeters: 167(W) x 58(H) x 200(D). Inches 6.56(W) x 2.28(H) x 7.88(D). Projections not included. |
| Weight | 2.5kg (5.5lb). |
| **TRANSMITTER** | |
| Output power | HF & 50MHz: 5 - 100W (AM 2 - 40W); 144MHz 1 - 10W (AM 1 - 4W). |
| Spurious emissions | HF - better than -50dB; 50 and 144MHz - better than -60dB. |
| Carrier suppression | Better than 40dB. |
| Unwanted sideband suppression | Better than 50dB. |
| Microphone impedance | 600Ω. |
| **RECEIVER** | |
| SSB, CW, AM, RTTY, FM | Double conversion superheterodyne |
| WFM | Triple conversion superheterodyne |
| Intermediate frequencies (approx) | SSB, AM, CW, RTTY: 69MHz and 9MHz. FM: 69MHz, 9MHz and 455kHz. WFM: 70.7 and 10.7MHz. |
| Sensitivity with pre-amp on (* not guaranteed outside amateur bands) | SSB, CW (for 10dB S/N) 1.8 - 29.9950MHz*, 50 - 54MHz, 144 - 148MHz: less than 0.16μV. AM (for 10dB S/N) 0.5 - 1.8MHz less than 13.0μV; 1.8 - 29.9950MHz*, 50 - 54MHz, 144 - 148MHz less than 2.0μV. FM (for 12dB SINAD) 28.0 - 29.7MHz less than 0.5μV; 50 - 54MHz, 144 - 148MHz less than 0.3μV. WFM (for 12dB SINAD) less than 10.0μV. |
| Squelch sensitivity threshold(pre-amp on) | SSB less than 5.6μV, FM less than 0.3μV. |
| Selectivity | SSB/CW >2.3kHz (-6dB), <4.0kHz (-60dB); AM >6.0kHz (-6dB), <20.0kHz (-40dB); FM >12.0kHz (-6dB), <30.0kHz (-50dB); FM narrow >8.0kHz (-6dB). |
| Spurious and image rejection ratio | More than 70dB (HF bands only). |

# Alinco DJ-G5 Dual Band Handy

***Review by RSGB HQ Staff***

ALINCO HAS A reputation for producing first class hand-helds and the DJ-G5 is no exception. Designed to replace the very popular DJ-580, it is smaller and covers the 144MHz and 430MHz bands. It is capable of running up to 5W on each band (with an external supply), and provides a huge range of 'bells and whistles'.

The radio comes with a helical antenna, a NiCad battery pack (4.8V), belt clip, hand strap, battery charger, instruction manual and a credit card sized aide-memoir for the most used controls.

The two bands are described as left (L) and right (R), corresponding to the display position. Each band can be designated Main or Sub and the priority of a number of the radio's operations depends on whether the Main or Sub band is selected. It is possible to use either band, simply by switching between the two, or receive on both bands or transmit on one and receive on the other (full or half duplex).

Four VFOs are provided, two on each band, plus up to 107 memories per band. Six memories are used for programming scanning limits, and one is a Call channel. The memories will contain receive frequency, split, transmit frequency, tuning step, tone setting, tone frequency and DSQ information (see below). Memory contents can easily be transferred to any of the VFOs.

## DESCRIPTION

THE DJ-G5 FITS comfortably into the hand and has a good solid feel to it. Its size is just 57W x 138H x 27Dmm, plus the knobs and antenna. It weighs in at just 350gm, complete with antenna, belt clip and the supplied battery pack.

The front panel is absolutely full. It contains no less than 20 buttons, almost all of which carry out two or three roles dependent on whether a function button has previously been pressed. The buttons include ON/OFF, band switching, frequency input, shift, scan etc. Two LEDs indicate an open squelch or transmitter on. There is a speaker close to the top of the front-panel and a microphone at the bottom - the radio can be held just like a portable telephone if required.

Pressing the buttons produces a variety of 'beeps' which can no doubt help to confirm an action in poor light or when initially setting up the radio. After a while this facility can become annoying and it can easily be switched off.The front-panel buttons can be disabled if required, as can the frequency knob on the top.

A quarter of the front is occupied by the LCD display which can be backlit for use at night. This includes two displays of frequency, repeater shift direction (+ or -), transmitter output setting (L, M or H), and tone squelch status. Displays common to both bands (or applying only to the Main Band) include a bargraph, battery status, auto power off, battery save, and whether duplex operation has been selected.

One side of the DJ-G5 carries the 'function' button, the main push-to-talk switch, a programmable push to talk button, and a MONI button which temporarily opens the squelch. Also on this side are the controls for volume and squelch which look at first glance to be thumbwheels but are actually UP/DOWN buttons linked by a piece of rubber. These are comfortable to use and facilitate single-handed operation.

On the other side is just a DC IN socket covered by a rubber plug.

In addition to a BNC aerial socket, the top has speaker and mic sockets (again covered with a rubber plug) and a rotary knob. This knob, which has a positive 'clicky' feel, is a good size and performs several user-programmable functions, including tuning, so if you hate tuning by UP/DOWN buttons this rig is for you. Tuning can be in 5, 10, 12.5, 15, 20, 25, 30 or 50kHz steps. Direct frequency entry is still available from the keyboard, of course.

Repeater operation is straightforward once programmed. A single RPT button selects repeater operation and pressing REV switches the receiver to the transmit frequency. Initial setting up allows any repeater shift in

100kHz steps up to 15.9MHz in either direction and a wide selection of CTCSS tones.

Two functions save valuable battery power: Auto Power Off disables the rig after a programmable period of non-use; Battery Save cycles the radio on and off in a programmable ratio.

Raynet groups may like to use the cross band repeater option (under official permit, of course). This relays the output of the VHF receiver on a UHF frequency or vice-versa.

## TRANSMITTER

**WITH THE SUPPLIED battery, the power output is approximately 1.5W on 2m and 1.0W on 70cm. Larger, optional batteries will give up to 4.5W on each band, and an external 12V DC supply allows a full 5W to be achieved. It would be quite feasible, then, to use the DJ-G5 as an FM base station, powered from the shack 12V supply, when not in use portable. Lower power can be selected from the front panel and a bargraph shows relative power output.**

The second push-to-talk switch, just below the main one, can be programmed to transmit at low power, to precede the transmission with a 1750Hz tone or to transmit on the Sub band.

CTCSS tones are supplied as standard and may be selected in VFO Mode. However, the reviewer was unable to store these in Memory Mode, though this is clearly a design facility. Also supplied is DSQ Code, a DTMF selective calling system. DTMF tones can be output manually, too. An Auto Dialer sends pre-set DTMF tones, presumably for phone patch which is not permissible in the UK.

## RECEIVER

**THE MOST NOTICEABLE thing about this receiver is the length of time it can be used on a single charge of the supplied battery. Plainly, a small battery will support only a small amount of transmitting time, but too often there is only an hour or so of monitoring time, too. The receiver will function long after the BATT (re-charge due) indicator shows.**

The receiver is sensitive and has a stable squelch. As supplied, it can be tuned over the whole of the two amateur bands. Extended receive coverage to 108 - 174, 420 - 470 and 850 - 950MHz is available (the instructions for this are contained in a typed leaflet produced by Waters & Stanton) as is AM reception. Transmission is restricted to the amateur bands.

A Bell facility produces 'beeps' when a signal is received - could be useful if you are prone to leaving the volume down but, fortunately, this can be switched off. In case of overload by adjacent transmissions, a 15dB attenuator can be switched in - this seems a most useful facility.

Three types of scanning are possible: the whole band, between one of three sets of programmable limits and all programmed memories. Scan resume can be programmed for 5 seconds after finding a signal, or 2s after the signal disappears.

The bargraph produces an 'S' meter display or, more interestingly, Channel Scope - a graphical display of the strength of signals on 10 adjacent channels (either in VFO or Memory mode). This can be most useful when monitoring for activity on a quiet band, or for instantly locating a free channel to move to from the calling channel. Channel Scope will display VFO channels or memory channels, and monitors in real time without interrupting the audio from the frequency you are listening to. Sweep Scan is similar to Channel Scope but the display moves with the scanning functions described above.

Priority Watch allows a favourite channel - say the local net frequency - to be checked for 0.2 seconds every 5s and, if a signal is received an alarm sounds and the receiver stays on the net channel for 2s.

For selective receiving, CTCSS Tone Squelch can be enabled using any of 50 tone frequencies. Alternatively, a 'DSQ code' using DTMF can be programmed in so that only stations using your code will open the squelch.

The two receivers (L and R) can be used for monitoring two bands at the same time with independent volume and squelch settings. It is also possible to receive simultaneously on two parts of the same band.

## HANDBOOK

**IN KEEPING with the rig's size, the handbook is only 128 x 182mm (a little smaller than A5) so it could be carried in an overcoat**

pocket, perhaps. It comprises 72 pages, extensively cross-referenced, with copious diagrams of the front panel and display. The most basic functions are dealt with in the earlier chapters with the most complex at the end. The book could be clearer, given the complexity of the radio, and even the most experienced reviewer found difficulty in discovering how to program individual CTCSS tones into the memories.

A troubleshooting guide is included and advice is given on how to connect a TNC for packet operation.

## ACCESSORIES

OPTIONAL EXTRAS include a remote control microphone with UP/DOWN buttons; 9.6V NiCads to give 4.5W RF output on either band, or 4.8V NiCads with twice the capacity of the supplied battery (for longer transmit time); rapid chargers; a filtered 12V cigar-lighter lead and soft cases.

## CONCLUSION

THE DJ-G5 IS A fine piece of engineering, robust and sophisticated yet very small. It is ideal for the experienced operator who wants both 2m and 70cm in a compact handheld, plus a wealth of features.

It does require a good read of the handbook before very much operation can take place and a one page 'get you started' sheet would be handy.

If you can only afford one rig, the DJ-G5 can be run off a 12V shack supply to use as a fully-featured 5W dual-band base station, and with external microphone and loudspeaker, it will work mobile in a quiet car.

## AVAILABILITY

THE RADIO RETAILS at £479 and is available from Alinco dealers across the country. The distributors, Waters & Stanton Electronics of Hockley (see p66), are thanked for the loan of the review model.

# SPECIFICATIONS

Sources: DJ-G5 Handbook and Waters & Stanton Electronics

## GENERAL

| | |
|---|---|
| Modes | F2E, F3E (FM). |
| Antenna impedance | 50Ω. |
| Operating temperature range | -10°C to +60°C. |
| Supply voltage (rated voltage) | 4.5 to approx 16 VDC (rated at 13.8VDC external or 4.8VDC with internal NiCads). |
| Current consumption (regulated supply voltage) | Tx Hi output with 13.8V external or 9.6V or 7.2V NiCads: approx 1.4A on VHF and approx 1.5A on UHF.<br>Tx Hi with 4.8V NiCads: approx 1.0A on VHF and approx 1.2A on UHF.<br>Tx Mid with 4.8V NiCads: approx 0.8A.<br>Tx Lo with 4.8V NiCads: approx 0.5A.<br>Rx squelched (twin band): approx 85mA.<br>Rx squelched (single band): approx 50mA.<br>Rx with battery save on (4:1 ratio twin band): approx 25mA. |
| Ground | Negative |
| Dimensions (including projections) | 63W x 155H x 31.5DD mm. |
| Weight (including antenna, belt clip and supplied NiCad) | 350gm. |

## TRANSMITTER

| | |
|---|---|
| Frequency range | 144.000 - 145.995MHz; 430.000 - 439.995MHz. |
| Modulation | Variable reactance |
| Maximum deviation | ±5kHz. |
| Power output | With external 13.8VDC supply: approx 5W.<br>With 9.6V NiCad: approx 4.5W.<br>With 7.2V NiCad: approx 3.5W on VHF and 3.0W on UHF.<br>With 4.8V NiCad: approx 1.5W on VHF and 1.0W on UHF. |
| Spurious emissions | better than -60dB. |

## RECEIVER

| | |
|---|---|
| System | Double conversion superheterodyne |
| First IF | VHF - 38.9MHz; UHF - 45.1MHz |
| Second IF | 455kHz |
| Frequency ranges | 144.000 - 145.995MHz; 430.000 - 439.995MHz (can extend to: 108 - 174MHz; 420 - 470MHz and 850 - 950MHz). |
| Sensitivity | (spec guaranteed only within amateur bands)<br>144 -146MHz: better than -16dBμ when using Left band; better than -12dBμ when using Rightband.<br>430 - 440MHz: better than -12dBμ when using Left band; better than -15dBu when using Right band. |
| Squelch sensitivity | better than -20dBμ(0.1μV). |
| Selectivity (-6dB / -60dB) | more than 12kHz / less than 30kHz. |
| AF output at 10% distortion | 100mW into an 8Ω load. |

# TS-570D Kenwood DSP Transceiver

***Reviewed by RSGB HQ Staff***

KENWOOD'S LATEST HF offering, the TS-570D, is a 'medium-priced' transceiver incorporating DSP technology. Amongst its other main features, Kenwood claims 'a world first' - CW auto tuning.

### *FEATURES*

THE TS-570D includes all the HF bands from 1.8 to 28MHz, with a general coverage receiver providing receive capability from 500kHz to 30MHz. All modes - SSB, CW, AM, FM and FSK - are permitted. Maximum power output is 100W on all modes apart from AM, where it is 25W. The transceiver is medium-sized, measuring 270W x 96H x 271Dmm and weighing in at 6.8kg - this is considerably more than the TS-50 / DX-70 / IC-706 type of mobile transceiver, but still much smaller and lighter than the TS-950 / FT-1000 'base station' transceivers.

An automatic ATU is built in as standard - not an optional extra as on some other 'medium price' transceivers. A built-in electronic memory keyer, with three 50-character memories, also comes as standard equipment.

All the 'bells and whistles' which have now come to be expected as normal on modern transceivers are included, including 100 memory channels, scanning, IF shift tuning, switchable preamplifier, speech processor etc.

A dual-function keypad on the front panel allows direct frequency entry and many other functions, depending on whether or not the ENT (enter) button is pressed first.

Many of the operational parameters are designed to be set by the user and then left, or adjusted only occasionally. These are set up by means of a 'Menu' system which allows the user to tailor the TS-570D exactly to his or her requirements. No fewer than 47 different parameters, from dial brightness to voice equalization characteristics, can be adjusted in this way. However, Kenwood have realised that many amateurs share a transceiver with another member of the family, or that a transceiver may be used for top-level contesting at the weekends and 'rag-chewing' during the week, and that different parameters are required for different circumstances. They have therefore provided two Menus, A and B, which can be set up completely independently of each other, and then switched from one to the other at the touch of a button.

A button labelled PF (Program Function) allows the user to customise its use via the Menus in order to activate a function which is not supported by other front panel buttons, such as the optional voice synthesizer. Four more Program Function buttons are available on the optional MC-47 microphone (a Kenwood MC-43 fist microphone is supplied as standard).

A MULTI knob on the front panel is used for many different functions depending on the use of other push buttons, enabling the front panel to have a clear, uncluttered, look.

Two SO-239 antenna sockets are on the back panel, and a switch on the front panel allows the user to select the antenna in use.

The transceiver runs from 13.8V DC, it does not have a mains PSU built in. The maximum current requirement is 20.5A. A Kenwood PS-33 power supply is available as an optional extra, but this adds a further £245.95 to the price. Other options include the SP-23 external loudspeaker, and narrow CW and SSB IF filters.

### DSP

Kenwood's 16-bit DSP technology in the audio chain provides very flexible interference reduction and signal processing facilities. There are several different DSP functions. Firstly, DSP slope tuning (this is in addition to the 'standard' IF tune) which provides independent high and low-pass filtering on a concentric knob on SSB and AM. DSP Noise Reduction facilities are provided by a button which toggles between NR1 (for SSB), NR2 (for CW), and off. A BC (Beat Cancel) button eliminates heterodyne 'whistles' as soon as they come into the receiver passband.

## IN USE

BASIC RECEIVE OPERATION of the TS-570D was transparently simple. It was possible to receive signals, change bands and modes, go from 'ham band' to general coverage receive, swap VFOs, and even use some of the more exotic features such as the DSP Noise Reduction, without even opening the handbook. But then when you wish to transmit, you immediately realise that there is no microphone gain control and it is time to reach for the instruction manual!

Fortunately, Kenwood's operator's manuals have a reputation for being very clear, and the TS-570D manual is no exception. The MULTI knob is not only the microphone gain (after pressing the small MIC button) but also controls many other features. RF power output can be adjusted from 5W to 100W in 5W steps (5 to 25W on AM) using this knob, making the transceiver ideal for QRP enthusiasts. The same knob controls the keyer speed from 0 to 100WPM in 2WPM steps.

The keys, buttons and knobs are 'soft-touch' and respond with a satisfying 'click' so that you know they have been activated. The function of most of the controls and buttons is quite clear, although there are exceptions: it is not obvious what the push buttons labelled MIC or KEY do until you read the manual, and there are two buttons, with different functions, labelled CLR and CLEAR.

## RECEIVE

On receive, the TS-570D gave a good account of itself, with noticeably better front-end performance on the noisy 160, 80 and 40m bands in the evening than another (admittedly older and cheaper) transceiver. The DSP functions all worked well, but it must be realised that, as with any transceiver, digital signal processing in the audio chain (however sophisticated) is no substitute for IF filters. For example, anyone contemplating serious CW operation will still need to install one of the optional narrow CW filters, as the transceiver's AGC is affected by *all* signals in the IF passband (2.2kHz), even though you're only listening to a bandwidth of, say, 100Hz. It is possible to reduce the RF gain to combat the effect of other strong signals in the passband, but then you start to lose signals which are in the noise.

On SSB the DSP Noise Reduction facility worked as well as other DSP units tested. The effect is most noticeable when listening to very weak signals just in the noise. Pressing NR1 reduces the noise level, leaving the signals in the clear, but giving a curious 'drainpipe' effect to the audio. The effect is similar to the audio on a long-distance telephone call. NR2, recommended for CW, was less successful. It worked well on strong signals, but distorted weak ones. Yes, it removed the noise, but the result was uncomfortable to listen to.

The 'Beat Cancel' was the most impressive of the DSP interference-rejection facilities. Carriers from 'tuner-uppers' are nulled out automatically within milliseconds of them appearing. The filter is both deep (strong carriers are completely nulled out) and sharp (the audio of the required signal is hardly affected by the operation) and more than one carrier at a time can be removed.

The 'Auto CW Tune' facility alters the VFO frequency so that a received CW station is matched to the receiver pitch frequency. The frequency of the receiver pitch and transmitter sidetone are identical (and can be set between 400 and 1000Hz according to the user's wishes) so using Auto CW Tune should always ensure that you are zero beat with the station you call.

The optional SP-23 loudspeaker provided somewhat better quality audio than the built-in upward facing speaker, although the latter was perfectly adequate. Most amateurs will probably prefer using headphones for HF operation anyway.

## TRANSMIT

The transmitter provided a generous 100W output on all bands, and the built-in ATU worked well, matching a 20m dipole at both band edges in milliseconds, and taking perhaps three or four seconds to match more complex loads. As with other built-in automatic ATUs, it can be thought of as a 'line flattener' allowing the full 100W to be transferred to resonant antennas at the band edges, rather than a wide-range ASTU capable of matching long wires or other non-resonant antennas.

Both antenna sockets can be used for transmission, so if your antenna farm consists of a triband beam for 10, 15 and 20m and a trap dipole or G5RV for the other bands, both antennas can be permanently connected to the rig. The ANT 1 / ANT 2 setting is automatically stored with the current band, so each time you select the same band you get the correct antenna. The front panel selection of antennas also provides a very convenient means of carrying out 'A' / 'B' antenna comparisons, for example between verticals and dipoles on the low bands.

One disadvantage for those who operate mainly phone (and especially SSB) is that there is no way to monitor your own transmitted audio. This is surprising, as Kenwood have provided a speech processor and other means of tailoring the transmitted audio to the user's requirements, but no way of checking on the results.

Kenwood claim that the transceiver's improved heat dissipation characteristics allow it to deliver long transmission periods with improved reliability - useful both for 'ragchewers' and contesters.

## CONCLUSION

THE TS-570D replaces Kenwood's popular TS-450S / TS-450SAT and looks likely to become another best-seller. It has more features provided as standard than most other medium-priced radios. Indeed, what is, in fact, quite a simple transceiver to operate has just about every conceivable facility built in.

It seems likely that more and more new transceivers will offer built-in Digital Signal Processing (DSP) facilities as standard. Just recently, these facilities were available on top of the range transceivers only, but as the technology becomes more widely available, built-in DSP will probably eventually be seen even on 'budget' transceivers. We are not at that stage yet, but Kenwood are to be congratulated for incorporating DSP into this 'middle of the range' transceiver.

A few years ago, transceivers with these facilities were being sold at prices of around £3000. The TS-570D has all the facilities you would expect of a top of the range transceiver, but in a medium-price and medium size package. It therefore represents excellent value for money.

Many amateurs will already possess a suitable 20A or 30A mains power supply, but if they do not, the cost of one must be added to the price of the transceiver, unless it is to be used exclusively for mobile or portable operation.

The Kenwood TS-570D costs £1499.95 inc VAT from Kenwood dealers. Our thanks to Kenwood (UK) for allowing us to use the only model in the country for our review, and also for providing a PS-33 PSU and PS-23 external speaker. For a data sheet on the TS-570D, call Kenwood on 01923 816869.

## SPECIFICATIONS

*(abridged from TS-570D Instruction Manual)*

| | |
|---|---|
| **General** | |
| Mode | J3E (LSB, USB), A1A (CW), A3E (AM), F3E (FM), F1D (FSK) |
| Number of memory channels | 100 |
| Antenna impedance | 50Ω (with antenna tuner 16.7 - 150Ω) |
| Supply voltage | DC 13.8V ±15% |
| Current Transmit (max) | 20.5A |
| Current Receive (no signal) | 2A |
| Dimensions (projections included) | 281W x 107H x 314Dmm |
| Weight | Approx 6.8kg (15lbs) |
| **Transmitter** | |
| Frequency range | 1.8 - 2.0, 3.5 - 4.0, 7.0 - 7.3, 10.1 - 10.15, 14.0 - 14.35, 18.068 - 18.168, 21.0 - 21.45, 24.89 - 24.99, 28.0 - 29.7MHz |
| Output power | SSB, CW, FSK, FM<br>Max 100W, Min 5W<br>AM<br>Max 25W, Min 5W |
| Spurious emissions | -50dB or less |
| Carrier suppression | 40dB or more |
| Unwanted sideband suppression (modulation frequency 1.0kHz) | 40dB or more |
| Maximum frequency deviation (FM) | Wide ±5kHz or less, Narrow ±2.5kHz or less |
| XIT shift frequency range | ±9.99kHz |
| Microphone impedance | 8Ω |

REVIEWS

**Receiver**

| | |
|---|---|
| Circuit type | Double conversion (triple conversion on FM) superheterodyne. |
| Frequency range | 500kHz - 30MHz |
| Intermediate frequency | 1st: 73.05MHz; 2nd: 8.83MHz; 3rd: 455kHz (FM only) |
| Sensitivity | **SSB, CW, FSK (at 10dB (S+N)/N)**<br>500kHz - 1.705MHz<br>4μV or less<br>1.705MHz - 24.5MHz<br>0.2μV or less<br>24.5MHz - 30MHz<br>0.13μV or less<br>**AM (at 10 dB (S+N)/N)**<br>500kHz - 1.705MHz<br>31.6μV or less<br>1.705MHz - 24.5MHz<br>2μV or less<br>24.5MHz - 30MHz<br>1.3μV or less<br>**FM (at 12dB SINAD)**<br>28MHz - 30MHz<br>0.25μV or less |
| Selectivity | SSB, CW, FSK<br>-6dB: 2.2kHz, -60dB: 4.4kHz<br>AM<br>-6dB: 4kHz, -50dB: 20kHz<br>FM<br>-6dB: 12kHz, -50dB: 25kHz |
| Image rejection (1.8MHz - 30MHz) | 70dB or more |
| 1st IF rejection (1.8MHz - 30MHz) | 70dB or more |
| RIT shift frequency range | ±9.99kHz |
| Squelch sensitivity | **SSB, CW:** 500kHz - 1.705MHz<br>20μV or less<br>**FSK, AM:** 1.705MHz - 30MHz<br>2μV or less<br>**FM:** 28 - 30MHz<br>0.25μV or less |
| Audio output (8Ω, 10% distortion) | 1.5W or more |
| Audio output impedance | 8Ω |

# INDEX OF EQUIPMENT & SOFTWARE REVIEWS IN *RADCOM* - 1985 TO 1994

Note: Prior to 1989 RadCom pagination ran sequentially through the year.

# Index

**YAESU**



**ICOM**

**KENWOOD**



INDEX